R. A. Gregory, H. G. Wells

Honours Physiography

R. A. Gregory, H. G. Wells

Honours Physiography

ISBN/EAN: 9783742806796

Manufactured in Europe, USA, Canada, Australia, Japa

Cover: Foto ©berggeist007 / pixelio.de

Manufactured and distributed by brebook publishing software
(www.brebook.com)

R. A. Gregory, H. G. Wells

Honours Physiography

HONOURS
PHYSIOGRAPHY.

BY

R. A. GREGORY,

*Oxford University Extension Lecturer; Honours Medallist in Physiography;
Formerly Computer to Solar Physics Committee, the Royal College of Science,
South Kensington; Fellow of the Royal Astronomical Society; Foreign
Correspondent of the 'Revue Générale des Sciences;'*

AND

H. G. WELLS, B.Sc., Lond.,

*Lecturer in Geology at the University Tutorial College; Third in Honours in
Geology and Physical Geography at B.Sc.,; Fellow of the Zoological Society;
Fellow (in Honours) and Doreck Scholar of the College of Preceptors.*

LONDON :

JOSEPH HUGHES & CO.,

FROEBEL HOUSE, ST. ANDREW'S HILL,
DOCTORS' COMMONS, E.C.

—

1893.

PREFACE.

———♦———

WE would like in this preface to have both the design and the limitations of this work clearly expressed. It is essentially a supplementary book, and it is prepared confessedly for a particular examination. In the Elementary and the Advanced Physiography which have preceded this, the authors conceive that a comprehensive and fairly full account of the science of Physiography has been given. In many directions, however, there are questions admitting of much fuller treatment than is either possible or desirable in works essentially introductory. And this work largely resolves itself into expansions at this point and that, of matters already dealt with in these two elementary books. Under the circumstances absolute continuity of treatment has been found to be impossible, though we have done what we can to render the succession of subjects easy and their arrangement methodical.

The Honours Examination of the Science and Art Department in this subject is a discursive one. The boundary is especially without definition on the geological side. But the past questions of the examiners have by this time established a precedent. The astronomy is descriptive and not mathematical : the geology does not extend to palæontology or stratigraphy. Honours Physiography involves a sound knowledge of geological structures in general, and the forces moulding them, but not a knowledge of particular instances, save only so far as they illustrate the general laws. And to the astronomy and geology is further added a thorough knowledge of meteorology. The Honours candidate must above all be well acquainted with *recent* work. He should follow what is going on between the date of this book and his examination either by reading *Nature* or by watching the monthly summary of new work in physiography that appears in the *Practical Teacher*.

Examination questions are appended to each section. In some cases the reader will find no answer to some of the questions here. This is because the most recent results have already been incorporated with the Elementary or the Advanced book in this series.

In thus avowing their deliberate aim to meet examination requirements the authors have no doubt that they will lay themselves open to the cheap accusation of having produced a cram book. But it does not follow that a book written to carry students through an examination is a cram book, or that one not designed to that end is truly educational. Cram is the acquisition of knowledge without assimilation. It can only occur under the benignant rule of an examiner who does not understand how to ask questions, and whose conception of his function is a series of demands for lists and tables ready made, and for names and dates. Such an examination this Honours test in Physiography certainly is not. Whatever the motive of the candidate, if he is the shallowest hunter of certificates that ever lived, he has to take the fruits of not a little thought and much real and permanent knowledge into this examination, or go without the distinction he covets.

This indeed is one of the strongest arguments against that pious horror of examinations still in vogue in some circles. Here and there you may worry an exceptional man, or disappoint a genius for a year or two, by forcing his mind into the grooves of a syllabus ; but on the other hand, your average well meaning, energetic man is directed, and indeed forced, to heights quite above the reach of his unaided mediocrity. Undoubtedly the typical examination candidate asks of any particular knowledge, not, Is it worth knowing ? but, Will it get marks ? But an examiner who allows what gets marks to have any other boundaries than what is worth knowing, is, we sincerely believe, an exception. Professors Judd and Lockyer, who have practically determined the present form of the Honours stage in this subject, are not only pre-eminent authorities in their respective fields, but also teachers of wide experience and great reputation. The authors have had the benefit of instruction from these teachers in the past, and they feel no hesitation in expressing an opinion that in following the lines laid down by the Physiography questions, they have a guidance in writing such a book as this, far better than their own conceptions of the treatment of the matter would supply.

Finally, the authors wish to express their thanks to Mr. A. Fowler, Demonstrator of Astronomy at the Royal College of Science, South Kensington, for valuable aid in the preparation of the first chapter, and to Mr. A. M. Davies, Demonstrator in Geology at the Royal College of Science, and Mr. W. T. Pain for several useful suggestions. To Messrs. Swift, and Mr. L. Casella, thanks are also due for furnishing the blocks from which four of the illustrations are printed.

R. A. GREGORY.
H. G. WELLS.

CONTENTS.

HONOURS PHYSIOGRAPHY.

CHAPTER I.

ASTRONOMICAL INSTRUMENTS AND METHODS.

THE second chapter of *Advanced Physiography* deals with astronomical instruments and their use. Here we extend particularly those parts concerned with the adjustment of the transit instrument and equatorial. These two telescopes form the essential outfit of an astronomical observatory, hence we give a number of problems arising from their employment. A variety of examples on verniers are also given, because experience has shown us that students rarely understand thoroughly the principle involved in vernier construction.

Construction and Use of Verniers.—The rule for constructing a vernier is stated as follows:—' In general to read to the nth part of a scale division, n divisions of the vernier must equal $n + 1$ or $n - 1$ divisions on the limb, according as they read in opposite or in similar directions.' Suppose that the divisions on the scale of a certain instrument each represented 20 minutes of arc. Then to construct a vernier which, in such a case, will read to 20 seconds of arc, we must find the ratio of 20' to 20''. In 20 minutes of arc there are 1,200 seconds. Hence the ratio is 1,200 : 20, that is as 60 is to 1. The nth part of a scale division to which it is required to read, is therefore one-sixtieth. If the vernier has to read in the same direction as the scale, 60 divisions on it must be equal to 59, that is, $n - 1$, divisions on the limb of the instrument to which the vernier has to be applied. For a vernier to read in the opposite direction to the scale, the 60 vernier divisions would have to be equal to 61

B

on the limb. In either case the vernier would read to $\frac{1}{60}$ of $20'$, that is, $20''$. As another example of a similar kind, suppose it is required to construct a vernier reading to ten seconds of arc for an instrument having its scale divisions equal to 20 minutes of arc. We reason out the problem as follows:—Twenty minutes of arc are equal to 1,200 seconds. It is required to read to 10 seconds. But $10''$ are equal to $\frac{1}{120}$ of 1,200 seconds, and this fraction is the nth part of a scale division. Hence the vernier must be constructed so that 120 divisions upon it are equal to 119 or 121 divisions on the scale.

Vernier Problems.—The 'degree of accuracy' of a vernier is always equal to the length of a division on the scale of the instrument divided by the number of divisions of the vernier. Thus, if l represent the length of a scale division and n the number of divisions on the vernier, the degree of accuracy (d) is given by the formula:—

$$d = \frac{l}{n}$$

By means of this formula it is possible to solve any vernier problem, as will be seen by the following examples:—

Example 1.—In an ordinary barometer, the scale of which is divided into inches and tenths, a vernier is attached containing ten equal divisions. To what degree of accuracy can the height of the barometer be read?

In this case the length of a scale division is $\frac{1}{10}$ of an inch ($= l$) and $n = 10$.

Hence substituting in the formula

$$d = \frac{l}{n} \text{ we get}$$

$$d = \tfrac{1}{10} \text{ inch} \div 10 = \tfrac{1}{100} \text{ inch.}$$

Therefore the height of the barometer can be read to $\frac{1}{100}$ inch.

Example 2.—In a standard barometer an inch on the scale of the instrument is divided into 20 equal parts. Show how the vernier is divided which reads to $\frac{1}{500}$ of an inch.

Here $l = \frac{1}{20}$ inch, $d = \frac{1}{500}$ inch, and n has to be found. Using the above formula we get

$$\tfrac{1}{500} = \frac{\tfrac{1}{20}}{n}$$

Whence $n = \tfrac{1}{20} \div \tfrac{1}{500}$

$$= \tfrac{1}{25} \text{ inch.}$$

To construct this vernier, either 24 or 26 equal divisions are taken from the scale of the barometer as the length of vernier and divided into 25 equal parts. In general 24 parts are taken and divided into 25 equal parts because, as we have pointed out, in this case the vernier reads in the same direction as the scale on the instrument, whereas if 26 parts are taken as the length of the vernier, the vernier reads backwards.

Example 3.—The arc or limb of a sextant is divided into degrees and quarters of a degree (15′). If 59 scale divisions are taken as the length of the vernier and divided into 60 equal parts, how accurately will the vernier read?

Here $l = 15'$, $n = 60$, and d is required.

Substituting in the formula $d = \dfrac{l}{n}$

we get $d = \frac{15'}{60} = 15''$

Hence degree of accuracy $= 15''$

Example 4.—The best sextants have the arc divided at every 10′; show how the vernier is divided which reads to 10″.

Here $l = 10'$, $d = 10''$, and n is required.

We have, therefore, $10'' = \dfrac{10'}{n} = \frac{10}{60}'$

Hence $n = 60$, so that to construct this vernier, 59 scale divisions are taken as the length of the vernier and divided into 60 equal parts.

Example 5.—A graduated circle is divided at every 20′. Divide a vernier to read to 15″.

$l = 20'$, $d = 15''$, and n is unknown.

Using the formula, we have $15'' = \dfrac{20'}{n} = \frac{20}{80}'$.

Hence $n = 80$, therefore 79 scale divisions must be divided into 80 equal parts.

Example 6.—An arc of an instrument is divided at every 10′, and the vernier has 40 equal divisions; find the degree of accuracy.

$l = 10'$ and $n = 40$.

Therefore $d = \frac{10}{40}' = 15''$.

Adjustment of a Transit Instrument.—A transit instrument is of no use whatever unless it is accurately adjusted. It is essential that the instrument stand upon a firm base and also that there should be a clear view to the south. To roughly determine the direction in which the telescope should point, the sun can be watched rising higher and higher in the sky before noon and the instrument moved bodily so that the sun is as near as possible in the centre of the field of view of the telescope when the highest point is reached. This being done, the axis should be tested for horizontality. The determination is made by placing a striding level along the axis from one trunnion to the other. The levelling screws at the base of the instrument are then turned until the bubble of the spirit-level is central. If the axis is horizontal the bubble will also take a central position when the level is reversed end for end, presuming that the level itself is perfect.

To determine whether the telescope is at right angles to the axis, a distant object is observed and the direction in which the central cross wire cuts it is noted. The axis is then lifted out of the bearings and put back so that each extremity rests in the bearing previously occupied by the other. This reverses the telescope, and if the object when again observed is found to be cut in the same direction by the central cross-wire, the axis and the telescope are exactly at right angles. But if not, the frame carrying the cross-wires must be moved through half the observed error.

It is now necessary to adjust the instrument so that the telescope points absolutely north and south and therefore the axis lies east and west. One method of doing this is by observing a transit of the sun. The time at which the sun transits at Greenwich is found from an almanac and a watch is obtained that keeps Greenwich time. Now, if the observation were made on the meridian of Greenwich the time indicated by the watch when the sun transited would be the time of transit at mean noon tabulated in the almanacs. At places east of Greenwich, however, the sun transits before this time, and at places west of it the time of transit is later, the difference depending upon the longitude of the place. For every degree of longitude four minutes of time must be subtracted from the almanac time in order to obtain the time of transit for a place east of Greenwich and added for a place to the west of it.

Suppose the instrument to be set up in longitude 2° 30′ W., and the observer to be provided with a watch or chronometer indicating Greenwich time, and let the mean time of transit of the sun's centre given in the almanac for the day of observation be 11h. 56m. 8s. Adding the correction for longitude (2° 30′ = 10m.), the Greenwich mean time of transit at the place selected is found to be 11h. 56m. 8s. + 10m., that is, 12h. 6m. 8s. Two or three minutes before this time the observer takes his seat at the instrument, and a minute before he begins to count the seconds. Exactly when the watch indicates 12h. 6m. 8s. the central cross-wire should lie across the centre of the sun's image. If it does not, a correction in azimuth must be made by means of the proper adjusting screws, and the observation repeated on another day.

As we cannot observe the transit of the sun's centre it is usual to take the time of transit of each limb, the mean of the two then giving the time at which the centre crossed the meridian. If clouds only permit one limb to be observed, the observation can be reduced to the centre by correcting for the time required for the semi-diameter to pass the meridian, which is given in the *Nautical Almanac*.

A second method, and one that has the advantage of not requiring a knowledge of Greenwich time for its accurate performance, consists in observing the transits of two stars widely separated in altitude, but which transit at nearly the same time. The interval between the time at which the two stars ought to transit can be found from an almanac, and if the observed interval is equal to it the instrument is in perfect adjustment. If, however, the lower of the two stars is found to transit too soon, the telescope points to the east of the meridian, while a westward bias is indicated when the star transits later than it should have done.

A third method involves observations of the pole star, or other circumpolar star. The pole star can be observed both at its upper and lower culmination at intervals of 12 sidereal hours. If a transit instrument is in proper adjustment, the interval from lower to upper culmination will be equal to that from the upper to the lower. If the error of the instrument be to the east of north, the interval from lower to upper culmination will be longer than the other; if to the west, it will be the shorter of the two. In this way the transit instrument can be adjusted without a knowledge of the right ascension of any star, or of the exact time. All that is required is a timekeeper going at a uniform rate.

Relation between Mean and Sidereal Time.—There is a constant relation between mean and sidereal time: 365 mean solar days = 366 sidereal days; one mean solar day = 24h. 3m. 56·555s. of sidereal time, and one mean solar hour = 1h. 0m. 9·8s. sidereal time. Similarly, a sidereal day of 24 hours =

23h. 56m. 4s. of mean solar time, and one sidereal hour = oh. 59m. 50·5s. in mean solar time. Knowing these relations, either sidereal or mean solar time can be converted into the other by simple proportion. Suppose, for instance that it is required to find the sidereal equivalent for 6h. 30m. 25s. of mean solar time. We write down the proportion :—

1 (mean) : 6h. 30m. 25s. (mean) :: 1h. 0m. 9·8s. (sidereal) : Ans.

From this proportion we find the answer to be an interval of 6h. 31m. 8s. sidereal time.

Again, suppose it is required to convert the sidereal time 10h. 17m. 40s. into mean solar time. From the relation given above we obtain the following proportion :—

1 : 10h. 17m. 40s. :: 0h. 59m. 50·5s. : Ans.

And the answer is found to be an interval of 10h. 15m. 59s. mean time.

In order to save the trouble of these calculations, tables have been prepared which give the equivalent of any interval in either sidereal or mean time. Tables of this kind will be found in the *Nautical Almanac, Whitaker's Almanac,* and some astronomical text-books.

Conversion of Mean and Sidereal Times.—The sidereal time at any instant is the number of hours, minutes, and seconds that have elapsed since the first point of Aries crossed the meridian. At twelve o'clock (noon) Greenwich mean time on any day the sidereal time may be anything from 0 to 24 hours. In other words, the *sidereal time at mean noon* may vary between those limits. The sidereal time at Greenwich mean noon is given for every day in the year in the *Nautical Almanac.* If at 7h. 45m. on a certain day we wished to know the sidereal time at Greenwich, the answer would be obtained by looking out the sidereal time at mean noon from the almanac and adding to it the sidereal equivalent of 7h. 45m. mean solar time.

It sometimes happens that the sum of the two exceeds twenty-four hours, in which case 24 hours must be subtracted. In order to convert sidereal time into mean solar time, the interval of mean time equivalent to the sidereal time is found as before, by proportion or from tables, and then added to the mean time at Greenwich at the preceding sidereal noon. These conversions will be better understood by giving a few examples.

The sidereal time at Greenwich mean noon on May 5, 1892, was 4h. 57m. 30s. Find the sidereal time at Greenwich at 8h. 15m.

	h.	m.	s.
Sidereal time at mean noon	4	57	30
Sidereal equivalent of 8h. 15m. mean time	8	16	18
Therefore the required time is	13	13	48

The sidereal time at Greenwich noon on December 25, 1892, was 18h. 17m. 50s. Find the sidereal time at 9.40 G. M. T.

	h.	m.	s.
Sidereal time at mean noon	18	17	50
Sidereal equivalent of 9h. 40m.	9	41	35
Total	25	59	25
Subtract 24 hours	24	0	0
The sidereal time required is	1	59	25

The mean time of transit at Greenwich of the first point of Aries (sidereal noon) on January 1, was 5h. 15m. What is the Greenwich mean time at 8h. 13m. 15s. sidereal time?

	h.	m.	s.
Mean time at sidereal noon	5	15	0
Mean time equivalent of 8h. 13m. 15s.	8	11	54
The mean time required is	13	26	54

Longitude corrections.—Terrestrial longitude can be expressed either in degrees, minutes, and seconds of arc, or in hours, minutes, and seconds of time. In a sidereal day of 24 hours, any point on the earth's surface turns through 360° on account of the earth's rotation, so that 15° of longitude correspond to 1 hour of sidereal time, and one degree to four minutes. Knowing the sidereal time at any place, the sidereal time at the same instant at any other place with known longitude can readily be calculated, the correction being added for east and subtracted for west longitudes.

Example.—When the sidereal time at Greenwich is 3h. 45m. 10s., what is it at a place in longitude 7° W.?

	h.	m.	s.
Sidereal time at Greenwich . . .	3	45	10
Longitude correction (= 7 × 4 m.) .	−	28	0
Sidereal time at place in question .	3	17	10

When correcting mean times, the sidereal time equivalent of the longitude must be converted into mean time intervals by the method already given.

Example.—When the mean time at Greenwich is 7h. 30m., what is the local mean time at a place 15° E.?

	h.	m.	s.
Mean time at Greenwich	7	30	0
Longitude correction (1h. less 9·8s.)	0	59	50·2
Local mean time at place	8	29	50·2

The *Nautical Almanac* gives the sidereal time at mean noon at Greenwich for every day in the year, and from this we have seen that the sidereal time at any given instant of mean time can be readily calculated. Similarly the sidereal time at any other place at a given local time can be determined if the sidereal time at local mean noon be known. The latter can be deduced from that at Greenwich on the same day by simply applying a correction for longitude. Since an hour of mean time exceeds an hour of longitude by 9·8 seconds, the correction will be at the rate of 9·8 seconds for each hour of longitude, subtracting if east, and adding if west.

Example.—If the sidereal time at noon at Greenwich be 18h. 20m. 45s., what will be the sidereal time at local mean noon at a place in longitude 22° 30′ E.?

		h.	m.	s.	
22° 30′	=	1	30	0	Sidereal interval
	=	1	29	45·3	Mean time ,,
Correction	=			15·75	

	h.	m.	s.
Sid. time at mean noon at Greenwich = 18	20	45
Longitude correction =			15·7
Sid. time at local mean noon, long. 22° 30′ = 18	20	29·3

This correction must not be confused with the longitude correction which is made in finding the sidereal time at a place *at the same moment* that it is a given sidereal time at some other place. A more general problem in which the correction is introduced is the following :—

Example.—Given that the sidereal time at Greenwich at mean noon on a certain day was 13h. 54m. 20s., what would be the sidereal time at a place in longitude 75° W. at 8 p.m. local mean time?

$$75° = 5h.$$

∴ Acceleration of sidereal time $= 5 \times 9\cdot8s. = 49s.$

Sidereal time at local mean noon $= 13h. 54m. 20s. + 49s.$
$$= 13h. 55m. 9s.$$

Sidereal equivalent of 8h. mean time $= 8h. 1m. 18\cdot6s.$

Sidereal time at 8 p.m. local mean time $= 21h. 56m. 27\cdot6s.$

To Determine Greenwich Time and Clock Error.—
The accuracy of a timekeeper can be tested by means of a transit observation. The times at which important stars transit at Greenwich are given in various almanacs. After the correction has been applied for longitude, the result obtained ought to be indicated by the watch when the star was actually observed to transit. Sometimes, instead of the time of transit at Greenwich being given, the right ascension of a star and the sidereal time at mean noon are the known quantities. Right ascension signifies distance from the First Point of Aries reckoned along the equator, and is usually expressed in hours, minutes, and seconds of sidereal time. The sidereal time at Greenwich mean noon is the number of sidereal hours, minutes, and seconds that have elapsed since the First Point of Aries passed the Greenwich meridian. Here is an example of this kind of problem :—

According to a certain watch, the time of transit of a star over the central wire of a transit instrument fixed in longitude 6° 15′ E. was 6h. 49m. 54s. The right ascension of the star was 15h. 45m., and the sidereal time at mean noon on the day of observation was 8h. 30m. Find the error of the watch in Greenwich time (1 hour sidereal time = 59m. 50·2s. mean time).

The following is a solution of the problem :—

	h.	m.	s.
Right ascension of star (= sid. time of transit)	15	45	0
Sidereal time at local mean noon	8	30	0
Sidereal interval between local noon and transit	7	15	0
Correction for 6° 15′ E. long., *minus* ...		25	0
Sidereal interval from Greenwich noon	6	50	0
Therefore Greenwich mean time of transit is	6	48	52
Time indicated by watch	6	49	54
Therefore the watch is fast by		1	2

If the mean time at which a star transits is observed, and the difference of longitude between the place of observation and the Greenwich meridian is known, the mean time at which the star transits at Greenwich can easily be calculated. When, however, the sidereal time of transit is observed and the Greenwich mean time of transit is required, the problem is somewhat longer. Here is an example of the former kind :—

At a place in longitude 3° 20′ W. Vega was observed to transit at 9h. 30m. 52s. (G. M. T.) mean time. Find the time at which the star transits at Greenwich.

The place is west of Greenwich, therefore the star does not transit there until after it has passed the Greenwich meridian. The difference of longitude is 3° 20′, and the equivalent difference of time is 13m. 20s. Therefore Vega transits at Greenwich 13m. 20s. before it does so at the place of observation, that is, at 9h. 17m. 32s. This simple problem is expressed in a convenient form as follows :—

	h.	m.	s.
Mean time of transit at place	9	30	52
Correction for longitude, *minus*		13	20
Mean time of transit at Greenwich	9	17	32

An example of the second kind is as follows :—

At a place in longitude 6° 28′ E. the sidereal time of transit of a certain star was found to be 12h. 10m. 30s. The sidereal time at Greenwich at mean noon on the same day was 5h. 20m. 45s. Find the Greenwich mean time of the star's transit (1 sidereal hour = oh. 59m. 50·2s. mean time).

	h.	m.	s.
Sidereal time of transit	12	10	30
Correction for E. long. (in sidereal time) ...		25	47
Sidereal time at Greenwich	11	44	43
Sidereal time at Greenwich noon ...	5	20	45
Sidereal interval between noon and transit...	6	23	58
Equivalent interval in mean time, that is, the			
G. M. T. of transit	6	22	54

This is the time, therefore, that a clock or watch keeping Greenwich time should indicate when the star in question passed the central cross wire of the transit instrument.

Determination of right ascension.—The right ascension of a star can be determined in a manner very similar to the above. Let us take a problem as an example of the calculation involved :—

On a certain day a particular star was observed to transit at 7h. 30m. Greenwich mean time at a place in longitude 2° 15′ W. The sidereal time at mean noon on that day was 3h. 45m. 30s. Given 1 hour mean time = 1h. om. 9·8s. sidereal time, find the R. A. of the star.

	h.	m.	s.
Mean time of transit at place of observation	7	30	0
Sidereal equivalent	7	31	14
Correction for 2° 15′ longitude, *minus* ...		9	0
Sidereal interval from Greenwich noon ...	7	22	14
Sidereal time at Greenwich noon	3	45	30
R. A. of the star, that is, time-interval between it			
and the first point of Aries	11	7	44

The right ascension of the star is therefore 11h. 7m. 44s.

Adjustment of an Equatorial Telescope.—A telescope intended to be used for the observation of stars in any part of the sky must clearly be movable in two directions. For a small telescope there is usually a vertical axis and a horizontal axis, this being called an alt-azimuth mounting. In order to observe a star continuously with such a telescope it must be moved step by step on each axis, unless the observer happens to be either exactly on the equator or at the north or south pole.

This form of mounting is therefore not so convenient as that in which continuous observations may be made by giving one motion only to the telescope, especially as this may be given by a driving clock. When it is remembered that the principal apparent motions of the stars are due to the earth's rotation, it is not difficult to see that they will appear to move round the earth's axis, and that the best mounting for a telescope will be that which allows of it being moved round a similar axis. Hence, in an equatorially-mounted telescope, one of the axes is put in such a position that it is parallel to the earth's axis; or since the earth's radius is infinitesimal in relation to stellar distances, it may be regarded as coincident with the earth's axis itself. This principal axis of an equatorial telescope is called the *polar axis*, and to make it parallel to the earth's axis is the chief adjustment of an equatorial telescope.

The declination and hour circles should also be made to give correct readings. Further, the declination axis must be truly at right angles to the polar axis, and the telescope tube must be perpendicular to the declination axis, but these are usually correctly adjusted by the maker of the instrument.

The adjustments which it is necessary for us to consider then are :—

(1) The polar axis must have an inclination equal to the latitude of the place.

(2) The polar axis must lie in the plane of the meridian.

(3) The declination circle must read 0° when the telescope is pointing to the celestial equator.

(4) The hour circle must read oh. om. os. when the telescope is pointing to any point on the celestial meridian.

In erecting the instrument, the polar axis is first placed approximately in position by inclining it at an angle as nearly as possible equal to the latitude of the place of observation, and then turning it so that it is nearly in a north and south vertical plane, judging this either by the indications of a compass or by reference to the pole star. The circles are also roughly adjusted by estimation, and the final adjustments are then made by observing suitable stars.

First Adjustment.—A star near the meridian is first observed with the telescope and the reading of the declination circle taken. The instrument is then turned to the opposite side of the pillar and the same star re-observed and the declination reading taken again. As the two readings are taken in opposite directions from the zero of the circle, the mean of the two will give the declination as measured by the instrument; that is, the declination of the star reckoned from the plane at right angles to the polar axis of the instrument. If the polar axis be correctly inclined, this instrumental will be equal to the true declination, as taken from the *Nautical Almanac* or other star catalogue. A little thought will show that if the instrumental reading be lower than the true declination in the case of a northern star, the polar axis is inclined at too small an angle. For a southern star the instrumental reading would be the greater in this case. Thus, in Fig. 1, if the angle δ indicates the true north declination of the

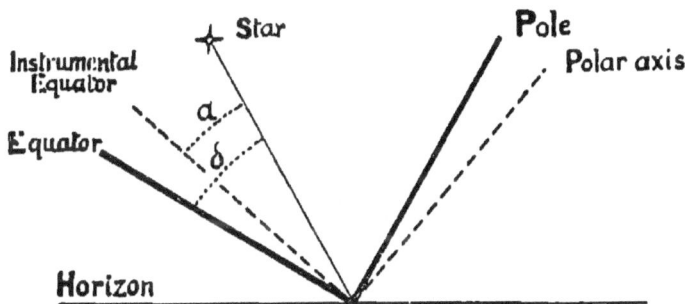

Fig. 1. Illustrating the adjustment for inclination of the Polar axis of an equatorial.

star, and the angle α the instrumental declination, it is clear that the polar axis must lie below the true pole. If the instrumental reading be too high, in the case of a northern star, the inclination of the axis is too great. A correction for atmospheric refraction has to be applied to each reading, and hence it is desirable that a star at a pretty high altitude be selected.

The following actual readings of α Hydræ, in latitude 14° N., will serve as an example :—

Declination reading with telescope east of pillar $= -6°\ 41'\ 20''$
　　　,,　　　　　　　,,　　　　west　　,,　$= -6°\ 55'\ 0''$
　　　　　　　　　　　　　　　　Mean　$= -6°\ 48'\ 10''$
Refraction correction (altitude $= 90° - 14° - 8°$
　　　　　　　　　　　$= 68°$)　$=$ _____ $23''$

　　　　Instrumental declination　　　　$= -6°\ 48'\ 33''$
　　　　True declination　　　　　　　$= -8°\ 11'\ 40''$
　　　　Error of polar axis　　　$= 1°\ 23'\ 7''$ too high.
　　This error is corrected by gradual approximations by mechanical
contrivances provided for the purpose.

Second Adjustment.—To put the polar axis in the plane
of the meridian, use is made of the declination circle, any error in
this direction largely affecting the declination readings of stars
away from the meridian. Fig. 2 will help to explain this in a
latitude such as that of London.

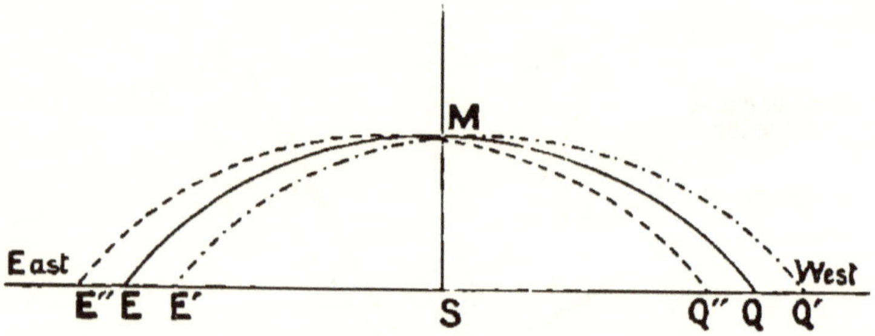

Fig. 2.　Illustrating the adjustment of the Polar axis of an equatorial
into the plane of the meridian.

　　The arc EMQ represents the portion of the celestial equator
which is above the horizon, MS representing the meridian, looking
south. The curve traced by the telescope when directed to an
equatorial star will be coincident with this when the polar axis is
properly adjusted. Assuming that the axis is not correctly
inclined to the horizon, the telescope will describe some track
such as E'MQ' if the upper end of the axis be to the east of the
true north. If the error be to the west, the track of the telescope
will be as E''MQ''. Practically this is determined by observing
a star about six hours east or west of the meridian, the error in
the declination reading being then at a maximum, as will be

understood from the diagram.   The axis is gradually adjusted in azimuth until the declinations of stars east and west as read on the instrument are equal to those given in the *Nautical Almanac*.

Another actual example will serve as a further illustration:—

Observed declination of α Coronæ 6 hours east $= +27° 12' 20''$
True declination $= +27° 4' 40''$

This shows that the polar axis is $7' 40''$ east of the true pole, as will readily be understood by a consideration of the diagram.

Third Adjustment.—The observations necessary for the first adjustment are also sufficient for the third.   It has been pointed out that the two readings are reckoned in opposite directions from the zero point, and if the circle were correctly adjusted they would be equal.   If they are not equal the circle must be shifted (by the screws provided) through an angle equal to half their difference; this is done in practice by making the circle read the mean of the two while the telescope is still pointing to the star.

Thus, in the first example,

Difference of readings $= 0° 13' 40''$
Error of circle ($= \frac{1}{2}$ diff) $= 6' 50''$

Fourth Adjustment.—Hour circles are of two forms, the simplest being a single graduated circle attached to the polar axis and provided with means of adjustment, and read by a fixed vernier.   The circle should indicate oh. om. os. when the telescope is in the meridian.   With the telescope in such a position, the declination axis will be horizontal, and the adjustment is therefore made by setting the declination axis with a level and then setting the circle at zero.

In another form of circle both the upper and lower edges are graduated.   The upper portion is read on the fixed vernier as before, while the lower is read by a vernier attached to the polar axis, the circle itself being independent of the polar axis.   The adjustment, however, is made in the same way as before, the only difference being that the two verniers are to be set exactly opposite to each other when the declination axis is horizontal.   This form of hour circle has the advantage of dispensing with any calculation of the hour angle, this being performed mechanically by turning the circle until the fixed vernier indicates the R. A. of the body to be observed and then turning the whole telescope until the other vernier reads the sidereal time of observation.   As the latter travels with the telescope it will become a sidereal clock for the time being.

The Wire Micrometer.—This telescopic accessory, of which a description is given in *Advanced Physiography*, page 40, is used for a large number of purposes, among which the measurement of the apparent diameters of the planets, the dimensions of lunar formations, the determination of the orbits of binary stars, and the determination of stellar parallax by the differential method may be specially mentioned.

The instrumental readings are quite arbitrary, as the pitch of the micrometer screw may be anything whatever. It therefore becomes necessary to determine the relation between the separation of the spider threads produced by one turn of the screw and the corresponding angular space traversed in the field of view of the telescope. The 'value' of the micrometer screw, as it is called, will depend upon the focal length of the telescope employed. Thus, in the case of a telescope in which the focal length of the object glass is 10 feet, a celestial object which would form an image 1 inch in diameter would have an apparent angular diameter of $\dfrac{360°}{2 \times 10 \times 12\pi} = 28' \, 39''$. If the micrometer screw used with such a telescope had 50 threads to the inch, its value would clearly be $\dfrac{28' \, 39''}{50} = 34''\cdot3$. The same micrometer when used with a telescope of double the focal length would have a value of $17''\cdot15$.

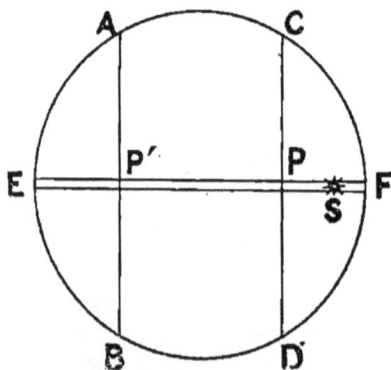

Fig. 3. The determination of the value of a micrometer screw.

In practice the value of the micrometer in relation to any particular telescope is determined as follows:—The movable wires AB and CD (Fig. 3) are separated by a known number of revolutions which we will indicate by the letter n. A star of known declination is then brought into the field of view and the whole micrometer is turned round until the fixed wires EF lie in the direction of the apparent motion of the star. The telescope being now directed so that the star falls between the fixed wires at S, the interval of time required for it to traverse from P to P' is determined, the telescope of course being at rest relatively to the earth, and the driving clock, if there be one, being disconnected. Let this interval = s seconds. The star observed will appear to travel at the rate of 15° per hour, as measured along a circle which is at right angles to the earth's axis, but since micrometric measures must be made on a great circle, this amount of arc must be converted into its equivalent on a great circle, that is, one whose plane passes through the earth. If δ indicate the declination of the star observed, the hourly motion measured along a great circle will be 15° × cos δ. In one minute the motion will be $15' \times \cos \delta$, and in one second $15'' \times \cos \delta$. In the case of a star on the celestial equator, cos δ = unity, and the apparent motion is at the rate of 15° per hour, such a star of course lying along the great circle of the equator. Bearing this time relation in mind, it will be seen that the value of the whole separation of the micrometer wires will be

$$s \times 15 \times \cos \delta \text{ seconds of arc,}$$

$$\text{or the value of 1 revolution} = \frac{s \times 15 \times \cos \delta}{n}$$

Example.—The wires of a micrometer were separated by twenty revolutions of the screw; α Coronæ was then observed to pass from one wire to the other in 17·5 seconds, the telescope being at rest and the apparent path of the star being perpendicular to the wires.

The declination of the star is 27° 4' 42'',

$$\text{Hence, value of 1 revolution} = \frac{17 \cdot 5 \times 15 \times \cos 27° 4' 42''}{20}$$

$$= 11''\cdot 68.$$

This value will be constant for the same instrument under similar conditions, but for work of extreme accuracy, temperature variations have to be taken into account, and each portion of the micrometer screw must be separately tested with a view to the detection of errors in the cutting of the thread.

(*a.*) To measure the diameter of a planet the wires AB and CD are made coincident with the opposite edges of the disc, and the number of revolutions separating the threads is read off. This multiplied by the micrometer constant will give the apparent diameter in arc. Thus with the instrument of which the value has already been found to be 11''·68, a measurement of the apparent diameter of Jupiter gave 3·4 revolutions; this is equivalent to 3·4 × 11''·68 = 39''·71.

C

(*b.*) To measure the distance between the components of a double star, the instrument is rotated so that the fixed wires EF lie along the line joining the two stars. The wire AB is then made to bisect one of the images and CD to bisect the other. The subsequent calculation is then similar to the last.

(*c.*) In determining stellar parallax by what is called the differential method, or proper motions of stars, the distances of the star under observation from two or three surrounding stars are measured in the same way as in the case of double stars. Any motion relative to the 'comparison stars' may thus be detected by repeated measurements.

The positions of comets are usually determined in the same way by reference to neighbouring stars of known position.

The Position Circle, which is usually attached to a micrometer intended for telescopic work, is used for determining the direction of the line joining any two stars, or of any other line, such as the direction of a comet's tail. The cross wires of the micrometer, in conjunction with the graduated position circle, enable this to be done very readily. To define a direction some

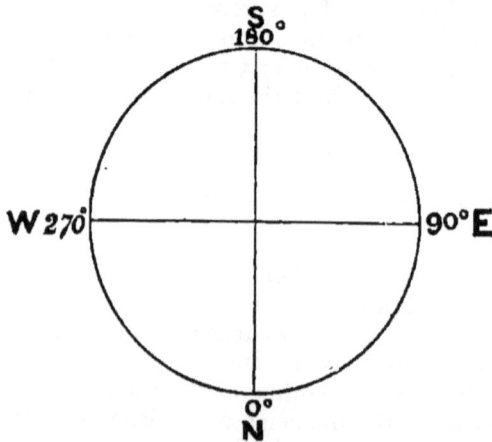

Fig. 4. The reading of position angles.

reference line is necessary, and the one selected is the meridional line passing through the north celestial pole and the point of the sky occupied by the object under observation. Position angles are then read from 0° to 360° from the north point through the east, south, and west points. Fig. 4 therefore represents the field of view for an inverting telescope.

Before measuring position angles, it is necessary to set the zero of the circle to the north point, or at least to determine the index error. In making this adjustment, advantage is taken of the fact that the apparent path of a star is along an east and west line. Hence the vernier by which the circle is read is first set to zero, and without disturbing this the whole instrument is turned round in the telescope until a star appears to traverse the direction of the fixed wires when the telescope is at rest. The wires at right angles will then be north and south, and the circle will be in adjustment.

Having set the instrument in its proper place with regard to the north point, the frame carrying the cross wires is rotated by a thumb screw until the zero wires lie along the line the direction of which is to be measured. The number of degrees read off on the circle will then indicate the angle through which the wires have been turned, reckoning from the north.

The position circle is chiefly used in the observations of binary stars, these measurements combined with the micrometric measures enabling the orbits to be deduced.

Determination of the Orbit of a Double Star.—At least five measures must be made, by means of a micrometer, of the distance between the two components of a binary and the position angle of the line connecting them. Generally a large number of determinations of this character are recorded, extending over several years. If the plane of the orbit of the double star is perpendicular to the line of sight, each star will be seen to describe equal areas in equal time round the centre of mass, and the period can be calculated when an arc of the elliptic orbit is known. The distance between the components in miles can only be found when the parallax of the binary has been determined. It is then possible to calculate the masses of the stars by the application of Kepler's third law. In practice, the larger component of a binary is supposed to be augmented by the mass of its companion, and the companion itself is considered as a particle revolving round this mass at a distance equal to the sum of the two semi-major axes of the apparent orbits. By this means only one orbit is found, instead of the two which result from a determination of the motions round the centre of mass. Usually the case is not so simple as that given. The orbits of binary stars are inclined in all directions to the line of sight, and what is really observed is a projection of the system upon the celestial sphere. In consequence

of this, the axes of the elliptic orbit may not appear at right angles
to each other and may not be the longest and shortest diameters.
What is more, even when the smaller star is considered as a
particle revolving round the larger, the latter body may not
occupy the focus of the elliptic orbit. In all cases, however, equal
areas are described in equal times. From this circumstance it is
possible to estimate the elements of the orbit of the binary with
more or less accuracy.

The Setting of an Equatorial on a Given Object.—If
the polar axis and circles be correctly adjusted, this is a very simple
matter. In many instruments the mechanical arrangements are
such that the telescope can only be used on one side of the
meridian without reversing it to the opposite side of the pillar.
That is, when working in the east, the telescope must be on the
western side of the pillar, and *vice versâ*. The first thing to be
done in setting the telescope to a given object is therefore to
determine whether it is east or west of the meridian. This
obviously involves the consideration of the time of observation,
since during one-half of the day any particular object will be east,

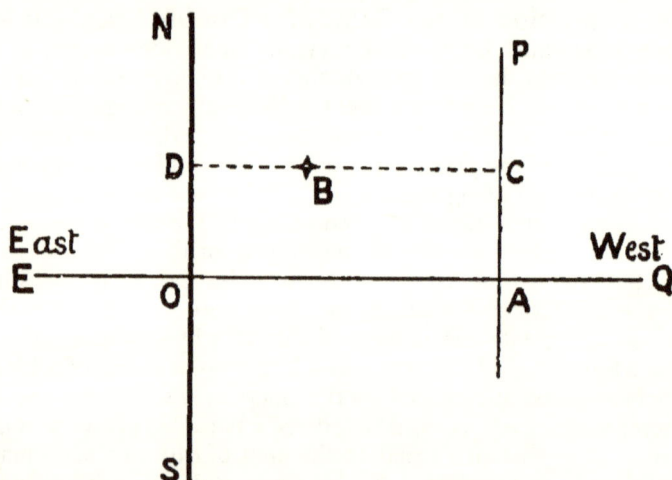

Fig. 5. Relation between hour angle, sidereal time, and right ascension.

and during the other half west. If no sidereal clock be at hand
it will be necessary to calculate the sidereal time at the time of
observation by the rules given on pages 7 and 8. The sidereal time
indicates the number of hours, minutes, and seconds which have

elapsed since the first point of Aries crossed the meridian, and since this is the start-point of right ascensions, which are reckoned from west to east, the difference between the sidereal time and the R. A. will indicate the hour angle of the body at the time of observation, *i.e.*, the distance from the meridian reckoned in time. If the R. A. be a smaller quantity than the sidereal time, the object must be west of the meridian; if greater, it must be east. Fig. 5 may make this clearer. NOS represents the meridian of the place of observation; EQ the celestial equator; PA the celestial meridian passing through the first point of Aries, intersecting the equator at the point A, which, of course, indicates the first point of Aries itself. The distance AO is then a geometrical representation of the sidereal time. If B indicates the body to be observed, BC represents its R. A. (in a case where this is smaller than the sidereal time), and BD its hour angle. Since BD = OA — BC,

$$\text{hour angle} = \text{sidereal time} - \text{R. A.}$$

In this case the body B must clearly be west of the meridian. If the R.A. were greater than the sidereal time, the point B must fall to the eastern side of NS.

In this way, then, it is determined whether the body to be observed is east or west of the meridian and by what amount. The telescope is accordingly placed on the proper side of the pillar, and the hour circle made to read the *hour angle* of the body at the time of observation.

When so placed, it is clamped, and thus put in connection with the driving clock. If the telescope be now turned on the declination axis it will pass through all stars having the same R. A., but to observe the particular body in question it must be turned to such a position that the proper declination is indicated on the declination circle.

Example.—Set the telescope to observe the planet Neptune, at Greenwich Observatory at midnight on December 31, 1892.

		h.	m.	s.
Sidereal time at mean noon December 31	=	18	45	26
Sidereal equivalent of 12h. mean time	=	12	1	58
Sidereal time at Greenwich at midnight	=	6	47	24
R. A. of Neptune (from *Naut. Alm.*)	=	4	30	36
· Hour angle west	=	2	16	48

Declination of planet, Dec. 31 (from *N.A.*) = $20° \, 15' \, 10''$.

The two latter readings are accordingly set on their respective circles. When the hour circle is a double one, as already described, the R. A. is set on one edge, and the sidereal time on the other, the subtraction given above being thus performed mechanically.

It should be pointed out that a telescope only requires setting in this way for objects which are not visible to the naked eye, or to which the finder cannot be readily directed from a knowledge of the position of the body in relation to neighbouring stars, or for objects which are invisible in the finder.

Determination of the Velocity of Light.— Professor Simon Newcomb published, in 1885, a memoir containing a full account of the various methods which have been used from time to time to determine the velocity with which light travels, and giving his own experiments on the subject. The method he adopted is the same as that employed by Foucault. Light from a fixed source falls upon a mirror capable of being rotated, and is reflected to a fixed concave mirror. The latter reflects the beam back to the movable mirror, whence it passes to the original source. If the movable mirror be set in rapid rotation it will turn through a small angle while light travels from it to the concave mirror and back again. The reflected beam will therefore not coincide with the incident beam, in other words an image of the light source will be seen separate from the source itself, and the separation will depend upon the velocity of light, the rate of rotation of the mirror, and the distance between the mirrors. Hence, in experimentally determining the velocity of light by this method, increased separation can be obtained by increasing the rate of rotation of the mirror, or increasing the distance between the two mirrors. For mechanical reasons, the rotational velocity cannot be carried beyond a certain point, and if the distance between the mirrors is considerably increased, the image becomes too faint to be accurately measured.

Professor Newcomb's arrangement was constituted as follows :-- A heliostat reflected a beam of light to a vertical slit fixed at one end of a telescope. The light passed down the tube to the object glass at the opposite end and then to the revolving mirror. This consisted of a rectangular steel prism, polished and nickel-plated, the vertical faces of which reflected the incident beam. Two turbines were connected with the prism, one at the top, the other at the bottom, and were arranged so that rotation in opposite

directions could be set up. The rate of rotation was recorded
with an electrical arrangement whereby every twenty-eighth rotation
was registered on a chronograph. Two mirrors, fixed side by
side, reflected the light received by them from the faces of the rotating mirror. A receiving telescope was placed below the sending
telescope, and was capable of moving in a horizontal plane to
catch the reflected beam. The amount of motion rendered
necessary by the deflection of the beam was measured accurately
by means of microscopes. The distance between the revolving
and fixed mirrors was 3,722 metres. After overcoming a number of
difficulties brought about by the rapidity with which the mirror
was rotated, Professor Newcomb was able to determine the
velocity of light with a probable error of about one ten-thousandth
part.

In making a determination, the slit was illuminated, and the
observing telescope was fixed upon some division on the graduated
scale on which it moved. The image of the slit may then appear
more deviated than the telescope from the original direction. To
bring it into the centre of the field, the turbine working in an
opposite direction at the opposite end of the mirror was set in
action. The telescope was then fixed on the other side of the
direction of the incident beam and the first turbine put out of
action. A deflection in the opposite direction was thus obtained,
and by again starting the first turbine, the motion of the second
was gradually retarded until the image of the slit again occupied
the centre of the field of view. Double the angle of deflection
was thus measured, and it was not necessary to determine the
position of the zero, which is a great advantage. Elaborate pre-
cautions were taken to cut down the intensity of the beam until
the spider-lines or cross-wires in the field of view were only just
illuminated, while the image of the slit was quite distinct. As a
result, the velocity of light was found to be 299,860 kilometres
per second, with a probable error of about 30 kilometres per
second.*

Spectra in relation to temperature.—The investigation of
spectra in relation to temperature has a most important bearing on
many questions of astronomical physics. At the present time it is
not possible to state exactly the number of degrees at which a

---

* See *Nature*, vol. xxxiv., pp. 29 & 170, 1886.

particular change occurs, but a good idea can be obtained of the relative temperatures associated with certain spectra.

The first point to be noticed is that at a low temperature, such as that of a Bunsen burner, only certain substances can give any visible spectrum. Thus, though sodium and calcium give characteristic visible radiations at this temperature, there are other substances, such as a piece of iron or platinum, which simply become red-hot and give a continuous spectrum. Volatility evidently therefore enters into the question. On passing from the Bunsen to the oxyhydrogen flame, the spectra of some of the more refractory substances, including iron, may be studied, and at the still higher temperature of the electric spark or arc, all substances are volatilised and give out their characteristic radiations.

The next point is that the spectra of many, if not all substances are very different under different conditions of observation. For instance, hydrogen at a reduced pressure enclosed in a Geissler tube and illuminated by an induced electric current gives a spectrum of lines, but if at a high pressure its spectrum is continuous. Oxygen, again, has at least three different spectra depending upon the conditions of pressure and temperature.

Some of the more important facts relating to these changes may be summed up as follows :—

(1.) The most common variation noticed is the passage from a banded or fluted spectrum to a line spectrum as the temperature is increased. Spectra observed at the temperature of the Bunsen or oxyhydrogen flame, for instance, consist largely of bands and flutings, while the same substances volatilised at the higher temperature of the electric arc or the electric spark give spectra consisting entirely of lines. Salts of manganese, iron, lead, chromium, barium, strontium, calcium, magnesium, and copper are good examples. In general, it can be stated that a fluted or banded spectrum is special to relatively low temperatures. Hence, when such flutings are seen in the spectra of celestial bodies it is fair to assume that the absorbing atmospheres of these bodies are relatively cool. Such is the case in stars resembling α Orionis, in which metallic absorption flutings are seen ; and in such stars as 152 Schjellerup, where the spectra indicate the fluted absorption of carbon.

Comets also occasionally exhibit similar changes from flutings to lines. When at a considerable distance from

perihelion, the great comet of 1882 gave a spectrum
consisting of carbon flutings, but as the light and heat of
the comet increased with approach to perihelion, bright
lines of sodium, iron, and other substances became
visible.

(2.) In other cases, the effect of increased temperature upon
the spectrum of a substance is chiefly to add to the
number of lines. Sodium, for example, at the temperature of the Bunsen burner gives the familiar D lines ; at
a higher temperature, even at that of the oxyhydrogen
flame, another double line in the red, one in the green,
and one in the blue make their appearance. Metallic
Iron, which gives no visible spectrum in the Bunsen,
shows a very small number of lines and one or two
flutings in the oxyhydrogen flame ; the electric arc brings
out many hundreds of lines, but gives no flutings.

(3.) Sometimes the lines in the spectrum of a substance show
remarkable changes of relative intensity as well as of
number, when observed at different temperatures. This
has been particularly studied by Prof. Lockyer and others
in the case of calcium, a substance which appears to
be as widely diffused throughout the universe as
hydrogen itself. The principal lines in this spectrum
are a blue line at wave-length 4226, and the Fraunhofer
lines H & K in the extreme violet. Fig. 6 illustrates

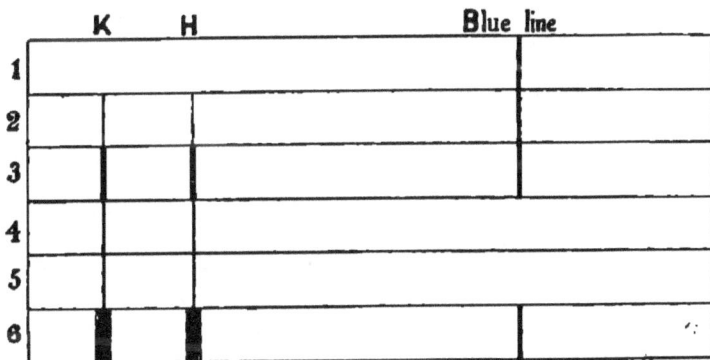

Fig. 6. Illustrating different spectra of calcium vapour. 1. In Bunsen
flame. 2. Oxyhydrogen flame. 3. Induction spark with small jar.
4. Large induction coil and jar. 5. Calcium spectrum in many of the
solar prominences. 6. Absorption of calcium in solar spectrum.

some of the different spectra, the diagram being
based on an illustration in Lockyer's *Studies in Spectrum
Analysis*.

Changes of this character in conjunction with other
lines of research, have led Prof. Lockyer to suggest
that the elements of the chemist are not really
elementary, for some elements are found to give different
spectra according to the degree of dissociation brought
about at any particular temperature. In the case of
calcium, the facts suggest that the blue line is produced
by the vibrations of a complex molecule, which is
broken up into finer molecules corresponding to the H
and K lines when a high temperature is employed.
This will perhaps be more clearly understood by considering what happens in the case of bodies which are
known to be compounds of the chemical elements.

(4.) In the case of some compound bodies there appears to be
a spectrum which is characteristic of the compound, as
distinguished from those of its component elements, but
in the majority of cases the compound molecules are
broken up at the lowest temperature which is sufficient
to render the radiations visible. When common salt is
burned in the Bunsen, for example, we do not see the
spectrum of sodium chloride, but simply that of sodium,
the same lines being observed whatever sodium compound be employed; in this case a very low temperature
is sufficient to dissociate the salt and give the spectrum
of sodium vapour.

Carbonic oxide furnishes a good illustration of the
changes which may take place in the spectrum of a
compound gas at different temperatures. For spectroscopic purposes a gas of this kind is enclosed at a low
pressure in a glass tube provided with platinum points,
so that an electric spark can be passed through it. Broadly
speaking, the varying intensity of the spark may be taken
to correspond to varying temperature. At a low temperature, the spectrum of the glowing gas includes certain
bands produced by the vibrations of the compound
molecules, these being characteristic of the compound;
mixed with this occurs a series of flutings of carbon
produced by partial dissociation. As the temperature is
increased, the bands of the compound disappear

altogether, leaving only the flutings of cool carbon. With increased temperature, another set of carbon flutings appear, and with a still further increase, bands and flutings disappear entirely and are replaced by the line spectra of carbon and oxygen.

Returning to the case of calcium, the blue line corresponds to the compound bands of carbonic oxide, while the lines H and K, which ultimately appear alone, would correspond to the spectra of the constituent elements of the possible compound which produces the blue line.

(5.) A knowledge of the relationships between spectrum and temperature gives valuable information relating to the conditions of the heavenly bodies. Reference has already been made to some of these points, but many others might have been added. For instance, the fact that the H and K lines appear without the blue line in the spectra of many of the solar prominences, is itself indicative of a very high temperature. Again, in stars which we have other reasons to suppose to be at a very high temperature, we get the H and K lines, with little or no trace of the blue line. In the cooler stars, on the other hand, such as α Orionis, the blue line is one of the most prominent in the whole spectrum. Indeed, it would be possible to arrange a very large number of stars in order of temperature by a consideration of the lines of calcium alone. But other features of the spectra usually point in the same direction, and the argument is strengthened. Thus, we see that minute researches in the laboratory are essential for a proper understanding of the spectra of the heavenly bodies.

QUESTIONS ON CHAPTER I.

1. Given an interval of 21h. 39m. 17s. in sidereal time, find the equivalent interval of mean solar time. (1 sidereal hour = 0h. 59m. 50·2s. mean time.)
Ans.—2h. 35m. 24s.

2. Find the interval of sidereal time that corresponds to an interval of 6h. 20m. 57s. mean solar time. (1 hour mean time = 1h. 0m. 9·8s. sidereal time.)
Ans.—6h. 21m. 59s.

3. The sidereal time at mean noon at Greenwich on a certain day being 20h. 12m. 30s., find the sidereal time at the same place at midnight.
Ans.—20h. 14m. 21·2s.

4. The mean time of sidereal noon at Greenwich being 8.30 p.m., what would be the mean time of the transit of a star of right ascension 5h. 20m.?
Ans.—13h. 49m. 7·7s.

5. When the sidereal time at Greenwich is 12h. 40m. 30s., what is it at places in longitude 35° E.?
Ans.—15h. 0m. 30s.

6. What is the local mean time at a place in longitude 18° E., when the mean time at Greenwich is 3 p.m.?
Ans.—4h. 11m. 48·2s.

7. The sidereal time at mean noon at Greenwich on a given day being 5h. 20m. 30s., what will be the sidereal time at a place 1° W. at 8 p.m. G. M. T. on the same day?
Ans.—13h. 17m. 48·6s.

8. At a place in longitude 2° W., α Orionis was observed to transit at 11h. 39m. 40s. p.m. by a clock intended to indicate Greenwich mean time. Given that the R. A. of the star is 5h. 49m. 20s., and that the sidereal time at Greenwich at mean noon on the same day was 18h. 17m. 20s., what was the error of the clock?
Ans.—1m. 34·7s. fast.

9. In the last problem what would be the error of the clock reckoned in local mean time?
Ans.—9m. 34·7s. fast.

10. What is the R. A. of a star which was observed to transit at 8 p.m. at Greenwich on Oct. 10, the sidereal time at mean noon on that day being 13h. 18m. 12s.?
Ans.—21h. 19m. 30·6s.

11. In the last problem what would be the R. A. if the observation were made in longitude 3° W., Greenwich time being kept at the place?
Ans.—21h. 31m. 30·6s.

12. How can the angular diameter of the sun or a planet be determined?

13. What are the chief uses of the transit instrument?

14. If you were given the right ascension and declination of a heavenly body, how would you proceed to point an equatorial telescope in the proper direction to observe the body?

15. How is a micrometer used to determine the angular diameter of a planet?

16. Enumerate the adjustments that must be made in order to put an equatorial telescope in working order.

17. How is the polar axis of an equatorial adjusted at an inclination equal to the latitude of the place in which it is erected?

18. How is the polar axis of an equatorial accurately adjusted in the plane of the meridian?

19. Describe the hour circle of an equatorial telescope, and the way in which it is adjusted.

20. Describe a wire micrometer and the method of determining the angular equivalent of each turn of the screw.

21. How would you measure the angular distance between the components of a binary star and the position angle of the line connecting them?

22. How is the orbit of a double star determined?

23. How is an equatorial telescope set to observe a heavenly body invisible to the naked eye?

24. Given a vernier reading to 20′, how is a vernier divided to read to 20″?

25. How would you divide a scale and a vernier in order that the vernier should read to 10″?

26. A certain circle is divided every 20′. How is a vernier divided so as to read to $\frac{1}{120}$ of 20′?

27. How is a vernier divided to read to 10″ when the scale of the instrument for which it is intended is divided every 20′?

28. How is a micrometer employed in the investigation of the parallax of a star?

29. Describe a method for accurately determining the position of a comet when it cannot be observed with a transit instrument.

30. What difference is most generally observed in the spectrum of a chemical substance at the temperature of the Bunsen burner and that of the electric spark?

31. Describe a laboratory experiment which suggests the dissociation of a chemical element.

32. What effect is produced in the spectrum of hydrogen by high pressure?

33. State briefly the varying spectra of metallic iron, at gradually increasing temperatures.

34. What is the laboratory evidence which enables us to discriminate between very hot and relatively very cool stars?

CHAPTER II.

STELLAR PHOTOGRAPHY AND CHEMISTRY.

THIS Chapter is supplementary to Chapters X. and XII. of *Advanced Physiography*. It contains a fuller description of some of the points there broached and also a few results and conclusions recently obtained in matters of stellar photography and spectroscopy.

The Chief Applications of Photography in Astronomical Observations.

(1.) *The Sun.*—Photographs of the sun are taken every day, and they furnish a permanent record of the state of the surface from year to year. The areas and positions of sunspots can thus be measured under the microscope at leisure, and, what is more, the measures can be verified at any future time. In addition to this series of sun-pictures, Dr. Janssen, of Meudon, takes photographs of sun-spots on a very large scale, and his pictures have extended considerably our knowledge of the structure of the sun's surface. During an eclipse, photographs are taken of the corona and prominences and their spectra; and by a special arrangement, Prof. Hale, of Chicago, is able to obtain photographs of the sun showing the spots, faculæ, and prominences upon it at the time of exposure. The solar spectrum is also photographed in juxtaposition with the spectra of different terrestrial elements, thus facilitating the determination of the chemical constitution of the sun.

(2.)   *Stars and Nebulæ.*—A photographic map of the stars is now being made by means of telescopes similar to that shown in Fig. 7. When it is finished, astronomers will have the positions of about 20,000,000 stars plotted on the

Fig. 7.   The Photographic Telescope of the Paris Observatory.

photographs. The fine details in nebulæ, too faint to appeal to the eye, have their light accumulated by the photographic plate until it leaves an impression. On account of this cumulative property, stars and nebulous material are photographed which can never be seen telescopically. The spectra of stars and nebulæ are also photographed. Prof. Pickering, of America, has catalogued 10,000 star-spectra. Prof. Lockyer has photographed the spectra of most of the brighter stars for the purpose of determining the order in which they have evolved. Prof. Vogel, of Potsdam, and M. Deslandres, of Paris, photograph star-spectra in juxtaposition with the spectra of terrestrial elements for the purpose of determining velocity in the line of sight. The late Prof. Pritchard has determined the parallaxes of stars by means of photography (see p. 34).

(3.)   *The Moon.*—Photographs of our satellite are taken from time to time and efforts are being made to obtain a continuous series, as in the case of the sun. Several observers have asserted that parts of the lunar surface have changed in recent years, and photography is undoubtedly the best means of settling such a question. Excellent lunar photographs have recently been taken at the Lick Observatory, California, by means of the 36-inch telescope. The moon's image in the focus of this immense instrument is more than five inches in diameter. The Brothers Henry, of the Paris Observatory, have also produced some marvellous pictures. By photographing the eclipsed moon and comparing the result with a photograph of the full moon, the difference between the amount of light received in each case has been found. Prof. Pritchard photographed the moon in order to investigate lunar librations.

(4.)   *Comets, Meteors, and Planets.*—Photographs of comets throw some light upon the structure of these bodies, but very little has yet been done in this direction. Dr. Max Wolf and others have photographed meteors as they flashed across the sky. From pictures of the same meteors seen from different places, the height of the atmosphere can be very accurately calculated and many important problems can be settled. A number of minor planets have also been discovered by photography. The photographs of major planets as yet taken are of little practical use except to verify a few visual observations.

Long-exposure Photographs of the Heavenly Bodies.
—A photographic plate is sensitive to rays of light, more especially to rays of short wave-length beyond the violet end of the spectrum. The more intense the actinic light, the quicker is an impression

produced upon the sensitive plate. When taking photographs of the sun or moon, the plate is only exposed for an instant to the light. Bright stars require a few minutes to leave their mark, fainter objects need a longer exposure. By increasing the time of exposure, fainter and fainter stars are grasped. The light which appears not to produce any effect after acting upon a plate for, say, five minutes, may have its existence revealed in double that time, and may form a distinct image in half an hour. It is this ability to accumulate faint light until it leaves its mark which gives the photographic plate a great advantage over the human retina. In recent years photographs have been taken of celestial objects with exposures of from one to four hours duration. By this means stars and faint nebulous matter have been revealed which it would be impossible for any observer to see with any telescope now in existence. Drs. Common and Roberts are the two English workers who have done most in this direction. Four photographs taken by the latter astronomer mark an epoch in the history of celestial photography. They represent (1) the Orion Nebula, (2) the Andromeda Nebula, (3) the Spiral Nebula in Canes Venatici, (4) the Pleiades. The Orion Nebula has been sketched and photographed by various astronomers. It is an excellent example of an irregular nebula. Dr. Roberts' photograph of the object (Fig. 8) shows that the nebulosity extends far beyond what was supposed to be the limit before the picture was obtained. Excellent drawings have been made of the Andromeda Nebula, but before Dr. Roberts' long-exposure photograph was taken, very little was known concerning the true form of this mass of celestial cloud. Observers had noticed two dark lanes or bands extending along one side of the bright central portion, but their significance was not understood. The photograph shows that these lanes are separations between rings of nebulous material which surrounds the bright central region like the rings surrounding the planet Saturn. To all appearances the rings have been left behind as the nebulous mass cooled and condensed. The Spiral Nebula in Canes Venatici was drawn in the form of a regular swirl or spiral by Lord Rosse, who used his six-foot reflector in the observation. Dr. Roberts' photograph shows that the regular stream depicted by Lord Rosse does not exist. The nebula appears to be like the nebula of Andromeda, but in a more advanced stage, for the rings surrounding the nucleus are knotted with brighter parts as if the material is condensing to form stars. But the most remarkable photograph ever taken is that of the Pleiades obtained by Dr. Roberts with four hours' exposure. To the naked eye, the Pleiades appear like other stars. Wisps of nebulous material surrounding some of the group were discovered photographically at Paris by the Brothers Henry. Dr. Roberts' picture shows that the stars are involved in a nebula. Streams or streaks of nebulous material stretch out from several of the stars, and have led Prof. Lockyer to conclude that the stars themselves

Fig. 8. The Great Nebula of Orion from a photograph by Dr. Roberts.

are really nothing but the intersection of a number of streams of meteorites. If two or more trains meet on a 'level-crossing' a considerable amount of heat is developed by the collision, and, according to Prof. Lockyer, this is what is happening in the Pleiades, with streams of meteorites substituted for railway trains. In 'The Meteoritic Hypothesis' he refers to this point as follows:—'The principal stars are not really stars at all; they are simply *loci* of inter-crossings of meteoritic streams, the velocities

D

of which have been sufficiently great to give us, as the result of collisions, a temperature approaching that of α Lyræ as far as we can judge by the spectrum. . . . . . . At the beginning of the action to which I have ascribed the present light of the Pleiades, we should have the appearance of a "new star," and the greater the light produced, and the more sudden the outburst, the more certainly would the appearance of the new star be chronicled.'

Stellar Parallax by means of Photography.—On account of the revolution of the earth round the sun, each star appears to describe a minute ellipse upon the celestial sphere in a year. The dimensions of each ellipse depends upon the distance of a star from the earth. Very accurate observations of position throughout the year enable the parallactic ellipse to be determined, but the best results are obtained by measuring the position of the star the parallax of which is required with respect to faint stars near it, and which are assumed to be so far away as to have no parallax—to be, in fact, reference points upon an infinitely distant sphere. If the assumption is true, the star under observation will be found to describe a minute ellipse with reference to the faint ones near it—a positive parallax is obtained. But if, on the other hand, the fainter stars are nearer the earth than the one whose parallax is being determined, a *negative* parallax is the result.

The late Prof. Pritchard successfully employed photography in the determination of stellar parallax. Photographs were taken from time to time of the selected star and the region near it, on small plates exposed in the principal focus of a reflecting telescope. Measures of the relative positions of the stars on the plates furnished the data from which parallax can be calculated. It was found to be unnecessary to continue the measures through a whole year. Photographs taken on twenty-five different nights were sufficient to give an accurate result. On each night four exposures were made, so a series of a hundred pictures of a star from twenty-five different points of view were obtained. Micrometrical measures of the relative positions of the images on the plates are certainly much easier and more conveniently made than in the telescope. An additional advantage lies in the fact that doubtful measures can be verified at any time, whereas, when an observation has been made in the telescope and recorded, it must stand on its merits, for the conditions can never again be exactly the same. Repeated trials have shown that the photographic film does not become appreciably distorted with the lapse of time, hence there is no doubt that the photographic method of determining stellar parallax will in the future be widely used. Prof. Pritchard compared results obtained photographically with those found by heliometer measures. In the case of 61 Cygni a parallax of $0''{\cdot}4294 \pm 0''{\cdot}0162$ was obtained. Bessel's probable error for the same star is practically the same as that here stated. The accuracy of the method is therefore placed beyond the possibility of doubt.

An examination of the results obtained leads to the conclusion that no relation exists between the lustre and parallax of stars, and, indeed, since stars are of all sizes and temperatures such a relation would hardly be expected.

Classification of Stellar Spectra.—Rutherford in 1863 pointed out that star spectra could be classified into different groups. He recognised numerous varieties of spectra, but considered all of them as modifications of three chief kinds. viz.: (1) spectra containing a large number of lines and bands, like the solar spectrum, (2) spectra wholly unlike that of the sun, (3) continuous spectra, that is, spectra containing no lines.

Secchi made a spectroscopic survey of the heavens, and as a result of his observations established four *types* of stellar spectra. The characteristics of the types are as follows :—

Type I.—Spectra given by white stars, like Sirius and Vega. Characterised by broad and dark lines, due to hydrogen. All other lines very faint or not visible at all.

Type II.—Spectra given by stars like Arcturus. Characterised by numerous fine lines resembling those seen in the solar spectrum.

Type III.—Spectra given by stars of a deep colour, like α Orionis and α Herculis. Characterised by dark bands or flutings, which fade away on their *less* refrangible sides.

Type IV.—This was added after Secchi's original classification. It contains a few stars like 152 Schjellerup, the spectra of which have bands like those in Type III., but fading away on their *more* refrangible sides.

It will be observed that the first two types have spectra made up of lines, while the third and fourth types are characterised by banded spectra. Prof. Pickering has suggested that a fifth type should be erected to include planetary nebulæ and certain bright-line stars.

In 1874 Vogel made a more elaborate classification than any previously put forward. His arrangement was based on the assumption that all stars at one time had a spectrum like that of Vega and that the deviations from this kind of spectra were brought about by decrease of temperature. The classification is as follows :—

Class I.—Spectra in which lines due to metals are extremely faint or quite invisible. The stars of this class are white. Their atmospheres consist almost entirely of hydrogen.

Class II.—Spectra in which lines due to metals are numerous and very distinct. Colour of the stars of this class, bluish-white to reddish-yellow. Hydrogen has given way to metallic lines.

Class III.—Spectra in which dark bands are seen, in addition to the lines due to metals. The stars of this class are orange or red.

Each of these classes is divided into sub-classes, so that all the peculiarities of stellar spectra are practically included in Vogel's arrangement. The one great objection to it is that it does not recognise the existence of stars increasing in temperature, but only those decreasing in temperature from the condition of Vega.

Professor Lockyer in 1887 suggested a new grouping of celestial bodies, in which those increasing in temperature are distinguished from those of decreasing temperatures. He recognises seven groups of stellar spectra, each of which is divided into species and sub-species. The last species of one group merges into the first species of the next above it, hence there is no hard and fast line of demarcation between any two groups. The criterion of each group is as follows :—

Group I.—Bright lines and flutings predominant. Includes nebulæ, bright-line stars and comets near aphelion. Dark flutings appear in the last species.

Group II.—Mixed bright and dark flutings. About 300 stars, many of them variable, are known to give this kind of spectra.

Group III.—Dark lines predominant. Evidence of *increasing* temperature.

Group IV.—Dark lines due to hydrogen predominant. Highest temperature. About one-half the visible stars belong to this group.

Group V.—Dark lines predominant. Evidence of *decreasing* temperature.

Group VI.—Dark flutings due to carbon predominant. About 60 stars are known to belong to this group.

Group VII.—Dark bodies having no spectra of their own.

Lockyer's Groups of Celestial Bodies.—The great difficulty to be overcome in order to establish the meteoritic grouping of celestial bodies is the determination of the criteria of temperature. According to Professor Lockyer's researches, when particles of a meteorite are gently heated in a vacuum tube, and an electric discharge passed over them, the element which first exhibits a spectrum is hydrogen. The spectrum of carbon is also occasionally seen. The first metal which shows itself is magnesium, a line due to this element appearing in the blue-green part of the spectrum. This line appears to occupy very nearly the same position as the chief line seen in the spectrum of nebulæ. On further increasing the temperature of the meteoritic particles, another line near that due to magnesium, and which also occurs in the nebular spectrum, becomes visible. The origin of this line is unknown. From these observations Professor Lockyer was led to conclude that the chief nebular line was due to magnesium. His suggestion was supported by the fact that the nebular line had been seen to have an undefined appearance on the more refrangible side, a peculiarity which also obtains to the magnesium line. When Dr. Huggins first observed the spectrum of a nebula he came to the conclusion that the chief line was coincident in position with a double line seen in the spectrum of nitrogen, and the nitrogen origin was implicitly believed in until the publication of Professor Lockyer's work. It is hardly too much to say that no spectroscopist now believes the line to be due to nitrogen. Indeed, Professor Young, one of the best American observers, says of the origin, 'It is now certain that, whatever it may be, nitrogen is not the substance.' Though something can be said in favour of the idea that the chief nebular line is due to magnesium, further observations have to be made before this origin can be considered to be established.

Professor Lockyer's researches have indicated that nebulæ are swarms of meteorites clashing and jostling one another, and so developing sufficient heat to volatilise the constituents which show themselves in the nebular spectrum. As the jostling increases, owing to the condensation of the swarm, the nebula increases in temperature until all the meteorites are driven into vapour. The nebula then becomes a star of the highest temperature in the heavens. Many of the substances we now know as terrestrial elements would probably be decomposed by the great heat, and there is reason to believe that hydrogen should be the prominent feature of a mass of such an elevated temperature.

The star then gradually cools down through various stages to the condition of the earth and moon. On this theory, therefore, some 'stars' should exist in almost the same condition as nebulæ, and others would be slightly hotter, and so on to the hottest stage. In other words, the meteoritic hypothesis recognises no hard and fast distinction between one class of celestial bodies and another.

Lockyer's Photographs of Stellar Spectra. — The method of obtaining photographs of stellar spectra by means of the 'slitless spectroscope' is described in *Advanced Physiography* p. 43. Recently several hundreds of stars have had their spectra photographed in this manner, under the direction of Prof. Lockyer at the South Kensington Observatory. The spectra obtained are two inches long and about one-eighth of an inch wide. In order to procure enlarged prints from these comparatively small negatives, a special device is employed. The negative of which an enlargement is desired is placed in front of an ordinary camera facing the sky, and steadily moved up and down by a clock-work arrangement in the direction of the lines in the spectrum. While this motion is going on, the negative is being photographed, and the resulting picture, instead of retaining the original proportion of width to length, is wider by the amount through which the negative moves up and down during the exposure. The lines in the spectrum are thus stretched out without any loss of light.

An investigation of the photographs taken and enlarged by these means gives support to Prof. Lockyer's hypothesis as to the mode in which celestial species are evolved, for a perfect agreement seems to exist between the phenomena to be expected on the hypothesis and the actual facts. The conclusions arrived at from the discussion are summarised as follows :—

(1.) *Nebulæ.*—The bright lines in the spectra of nebulæ are due to (*a*) hydrogen and some carbon, produced by the light which occupies the interspaces between the meteorites ; (*b*) magnesium, iron, and calcium at a comparatively low temperature, produced by the vapours driven off when meteorites graze one another ; (*c*) vapours at a temperature comparable to that of the sun's chromosphere, produced by the direct collisions of meteorites.

(2.) *Bright-line Stars.*—There is no hard-and-fast distinction between stars and nebulæ, some stars having a spectrum of bright lines precisely similar to the nebular spectrum.

(3.) *Stars of Increasing Temperature.*—The bright lines from the interspaces are neutralised by corresponding dark lines from individual meteorites. The bright bands of carbon then become predominant, while dark bands of manganese, lead, and iron appear. Further condensation of the meteor-swarm causes the carbon radiation from the interspaces to disappear, and dark lines replace the dark bands as a consequence of the increased temperature ; Alpha Tauri (Aldebaran) is a type of a star at this stage. The dark lines of many substances then disappear ; and since the chances of direct collisions are enormously increased, the absorption of vapours at a very high temperature is exhibited in the spectra of stars at this stage ; Alpha Cygni and Beta Orionis are examples of this type.

(4.) *The Hottest Stars.*—The lines seen bright in nebulæ, whatever their origins, appear almost alone as dark lines in stars of the highest temperature. Bright hydrogen is conspicuous in the nebular spectrum, and in the spectrum of the hottest stars dark hydrogen is the predominant feature ; Alpha Andromedæ is an example of this stage.

(5.) *Stars of Decreasing Temperature.*—As the depth of the absorbing atmosphere diminishes, the hydrogen lines become thinner and new lines appear. This is exemplified by the spectrum of Sirius. The hydrogen lines continue to thin out, and more dark lines of iron and other elements show themselves, as in Beta Arietis and Alpha Persei. With the further thinning out of the hydrogen lines and reduction of temperature of the atmosphere, carbon absorption commences. There is evidence that Arcturus and the sun are at this stage.

Motion of Stars and Nebulæ in the line of Sight.—

Numerous lines in the spectrum of terrestrial iron, hydrogen, magnesium, and other elements have been detected in stars. These lines have, under normal conditions, a fixed position or wave-length. If, however, a source of light is moving towards the earth, or the earth is hastening towards it, the wave-lengths of all the lines in the spectrum of the source will be increased. On the other hand, a motion of relative recession causes a decrease of wave-length, hence the lines are in this case shifted towards the red end of the spectrum. In order to determine whether the lines in the spectrum of a star have their normal wave-length

or not, the star-lines are matched by the same set of lines emitted
by the terrestrial element that they represent. Should each of
the star-lines appear on the blueward side of the fellow line of the
comparison spectrum, a motion of approach is indicated, while a
displacement to the redward side shows that the star and the
earth are in relative recession. The velocity of motion in the
line of sight can be calculated from the measured displacement
of the star-lines by means of the following formula—

$$v = \frac{d}{\lambda} V$$

where v is the velocity, λ the true wave length of a line, d the
difference between the true and the apparent wave-lengths, and
V the velocity of light. Dr. Huggins was the first astronomer to
apply this method in the determinaton of the motions of stars
in the line of sight, and he was followed by Mr. Christie, the
present Astronomer Royal, and Mr. Maunder at Greenwich
Observatory, and by Mr. Seabroke at Rugby. But the visual
observations made by these observers carried with them a large
margin of error, and it is only recently that any remarkable
degree of accuracy has been attained. The improvement is due
to the application of photography to the research. Professor Vogel,
at Potsdam, and M. Deslandres, of the Paris Observatory, have
obtained excellent photographs of star spectra side by side with the
spectra of terrestrial substances, such as iron and hydrogen. By
measuring the displacements of the celestial lines with regard to
their terrestrial fellows, the relative velocity in the line of sight of
the star under observation is found. In order to test the accuracy of
the photographic method, Professor Vogel photographed the spectrum of Venus in juxtaposition with a comparison spectrum. The
velocity of Venus relatively to the earth can be calculated, and a
comparison of the calculated velocity with the velocity determined
by means of the spectroscope shows a very close agreement
between the two. This brings us to the question as to the
influence of the earth's orbital motion on the determination of
stellar velocities in the line of sight. The point towards which
the earth is travelling is constantly changing. At one instant
the earth is moving in a particular direction, and six months later
it is moving in the opposite direction, having travelled to the
other side of its orbit. This being so, the apparent velocity in
the line of sight indicated by the displacements of the lines in the
spectrum of a star must be corrected by the velocity with which
the earth happens to be moving towards or away from the star at

the moment of observation. Professor Vogel claims that the photo-graphic method permits him to determine the velocities of stars in the line of sight with an accuracy of about a mile per second. He has published a list of fifty-one stars, the velocities of which have been measured at Potsdam. The list shows that Aldebaran has the maximum velocity of recession, viz., 30·2 miles per second, while the greatest velocity of approach is 24·0 miles per second possessed by γ Leonis. Seven stars are receding and eleven stars are approaching the earth with a greater velocity than 10·4 miles per second. Professor Keeler, by means of a spectroscope on the great telescope of the Lick Observatory, California, has been able to determine visually the motions of bright stars in the line of sight with an accuracy quite equal to that attained by Vogel. One example will suffice to show the agreement between the measures of the two observers. According to the Potsdam measures, Arcturus is moving towards the earth with a velocity of 4·4 miles per second. Professor Keeler's eye observations led him to assign the star a velocity of 4·3 miles per second; that is to say, the difference between the two results is only one-tenth of a mile per second. Professor Keeler has also determined the motions of some nebulæ in the line of sight. One nebula he found to be approaching the earth with a velocity of thirty-one miles per second, and another receding at the rate of thirty-eight miles. The great nebula in the constellation of Orion has a velocity of recession of about ten miles per second.

Spectroscopic Double Stars.—In 1889 Professor E. C. Pickering discovered that the K line in the spectrum of ζ Ursæ Majoris was doubled every fifty-two days. This periodical doubling is caused by the fact that the star really consists of two bodies of about equal size, so close to one another that no telescope can show their duplex character. If the component stars were at rest, their two spectra would overlap, but since they revolve in an orbit which can be likened to a hoop looked at edgeways, when one star is rushing towards us, the other is running away from us. The set of spectrum-lines belonging to the former body are therefore shifted slightly towards the blue end of the spectrum, that is, increased in pitch, while those of the latter experience a shift towards the red end, or are lowered in pitch. Hence, instead of overlapping, the two sets of lines are separated by a slight amount when the above conditions obtain. When, however, the two stars are moving *across* the line of sight.

that is, neither increasing nor decreasing their distance from the earth, there is no decrease or increase of pitch, and the two spectra overlap. Twice, then, in a revolution, are the spectrum-lines single, and twice are they duplicated. Fifty-two days elapse between two successive maximum separations of the lines, hence the period of revolution is twice this interval, that is, 104 days. Measurements of the amount of separation shows that the relative velocity of the components is about 100 miles per second. The stars are about 143 millions of miles apart, and their combined mass is about forty times greater than the mass of the sun.

A similar periodic doubling has been found in the spectrum of β Aurigæ, the intervals in this case being only two days, and therefore the period of revolution is about four days. The relative velocity is about 140 miles per second, that is, each star moves at the rate of seventy miles per second. The distance between the two stars is seven and a half millions of miles and their combined mass is 4·7 times the sun's mass. With regard to double stars of this character Dr. Huggins remarked, in his presidential address, delivered at the meeting of the British Association in 1891 :—' The telescope could never have revealed to us double stars of this order. In the case of β Aurigæ, combining Vogel's distance with Pritchard's recent determination of the star's parallax, the greatest angular separation of the stars seen from the earth would be one two-hundredth part of a second of arc, and therefore very far too small for the highest powers of the largest telescope. If we take the relation of aperture to separating power usually accepted, an object glass of about eighty feet in diameter would be needed to resolve this binary star. The spectroscope, which takes no note of distance, magnifies, so to speak, this minute angular separation four thousand times ; in other words, the doubling of the lines, which is the phenomenon that we have to observe, amounts to the easily measurable quantity of twenty seconds of arc.'

The System of Algol.—A century ago Goodricke suggested that the variability of Algol was caused by a dark body periodically passing between the star and the earth and partially eclipsing its light. Professor Pickering proved that the hypothesis fully accounted for the facts of observation, and Professor Vogel has placed it beyond the range of doubt. Twelve photographs of the spectrum of Algol, taken at Potsdam in comparison with a terrestrial spectrum, led to the discovery that the star swings towards and away from the earth in a period coincident with the variation of light, viz., 68·8 hours. Before minimum, the bright star is being swung back with a velocity of about 24 miles a second, while the dark companion is coming forward to obstruct its light. After a minimum, the spectroscope indicates that the bright star is approaching at the rate of nearly 29 miles a second, hence the dark star on the opposite side of the common centre of gravity is being swung back. From the difference between the velocities of approach and recession, it is found that the whole system of Algol is moving towards the earth with a velocity of 2·3 miles a second. By assuming that Algol and its dark companion have the same density, Professor Vogel concludes that the diameter of the bright star is 1,061,000 miles, and of the dark component 830,300 miles. The distance between the centres of the two bodies is 3,250,000 miles. The motion of Algol in its orbit is 26 miles a second, and that of the companion is 55 miles. Algol has a mass four-ninths the mass of the sun, and the mass of the dark body is two-ninths the same unit.

An investigation of the inequalities in the period of Algol's variation of light has led Dr. Chandler to assume the existence of a third dark body in order to explain them. He has stated his theory as follows in the *Astronomical Journal:*—' Algol, together with the close companion—whose revolution in 2d. 20·8h. produces by eclipse the observed fluctuations in light, according to the well-known hypothesis of Goodricke, confirmed by the elegant investigations of Vogel—is subject to still another orbital motion of a quite different kind. Both have a common revolution about a third body, a large, a distant, and dark companion or primary, in a period of about 130 years. The size of this orbit around the common centre of gravity is about equal to that of Uranus round the sun. The plane of the orbit is inclined about 20° to our line of vision. Algol transited the plane passing through the centre of gravity perpendicular to this line of vision in 1804 going outwards, and in 1869 coming inwards. Calling the first point the ascending node, the position angle, reckoned in the ordinary way, is about 65°. The orbit is sensibly circular, or of very moderate eccentricity. The longest diameter of the projected ellipse, measured on the face of the sky, is about $2''\cdot7$. A necessary consequence of this theory is an irregularity of proper motion with an amplitude of something over a fifth of a time second in right ascension, and nearly one and a half seconds in declination ; the middle point being the centre of gravity of Algol and the distant unknown companion, and the uniform proper

motion of the latter being — o·oo1os. and + o''·o12o annually
in the two co-ordinates (R. A. and Decl.) respectively. The
annual parallax of the star is about o''·o7. The mean period of
light variation is 2d. 2oh. 48m. 56·oos.'

Movements of Spica in the Line of Sight.

—In the
case of the Algol system, the orbit happens to lie nearly in the
plane of the ecliptic.   If the orbit were inclined so that the
dark companion pass above or below the bright star when on the
earthward side of it, evidently no eclipse of the star's light would
occur.   Nevertheless, such a star would be swung backwards and
forwards by the dark companion.   Spica appears to be a star of
this character—a body under the influence of an unseen world.
It swings backwards and forwards in a period of four days, each
component of the system moving at the rate of about fifty-six
miles per second.   In addition, the Spica system is moving
towards the earth with a velocity of rather more than nine miles
per second.   The mass of the system is 2·6 times the sun's mass,
and the distance between the two components is about 6¼ millions
of miles.

Composite Stellar Spectra.

—Professor E. C. Pickering
has suggested that certain stellar spectra are compound, that is,
formed by the integration of two or more kinds of spectra emitted
by as many different bodies.   When the components of a close
binary system have similar spectra, relative orbital motion in the
line of sight may cause a periodic doubling of the lines.   But if
the spectra be not similar, any lines common to both ought to be
conspicuously strong, and, provided the components have not
equal and opposite velocities in the line of sight, ought also to be
displaced with reference to other lines.   Thus, if one component
of a close binary system has a spectrum like our sun, and the
other a spectrum like Vega or Sirius, in which strongly marked
hydrogen is the main feature, the resulting spectrum will have a
composite character, and careful measurements should show that
the position of the hydrogen line is periodically displaced when
compared with the lines characteristic of the solar-type spectrum.
Alpha Canis Majoris is the brightest star having this composite
spectrum, and the wave-length of the hydrogen line G, derived
from a comparison with three lines of greater and three lines of
smaller wave-length, was 434·o9, which exceeds that derived from
the solar spectrum by o·o3.   From this displacement it would

appear that if the phenomenon is due to the relative motion of a faint component, the body is receding at the rate of twenty kilometres per second, as compared with the bright component. An examination of spectra has led Professor Pickering to decide that the spectra of about a dozen stars are of the compound character referred to. Although the strong hydrogen lines in the spectra investigated may be due to the presence of a faint companion, their intensity may also be due to many other causes. Thus, the strong hydrogen lines in the solar spectrum are not due to the integration of the spectrum of the sun and that of a companion. It is necessary therefore to determine whether the displacement is subject to a periodic variation or not in order to test this method of discovering close binaries.

New Stars and their Origin.—In the time of Tycho Brahe and Kepler 'new stars,' or Novæ, were supposed to be formed from the material of which the Milky Way was thought to consist. Newton considered that the phenomena were produced by the appulse of comets. Zöllner, in 1865, suggested that the bursting forth of luminous matter from the interior of a star covered with a dark crust, and the subsequent burning of the substances which formed the crust, would explain the creation of a new star. This volcanic theory was also favoured by Vogel and Lohse. Observations of Nova Coronæ in 1866 led Dr. Huggins to conclude that *'the star became suddenly enwrapt in burning hydrogen.'* To quote more fully, 'In consequence, it may be, of some great convulsion, of the precise nature of which it would be idle to speculate, enormous quantities of gas were set free. A large part of this gas consisted of hydrogen, which was burning about the star in combination with some other element. This flaming gas emitted the light represented by the spectrum of bright lines. The greatly increased brightness of the spectrum of the other part of the star's light may show that this fierce gaseous conflagration had heated to a more vivid incandescence the matter of the photosphere. As the free hydrogen became exhausted the flames gradually abated, the photosphere became less vivid, and the star waned down to its former brightness.' More or less modified forms of this theory of a fiery cataclysm were afterwards put forward to account for the formation of Nova Cygni in 1876. Professor Lockyer, however, advanced the idea that the outburst was due to cosmical collisions. In his words, 'We are driven from the idea that these phenomena are produced by the incandescence of large masses of matter because, if they were so produced, the running down of brilliancy would be exceedingly slow. Let us consider the case, then, on the supposition of small masses of matter. Where are we to find them ? The answer is easy : in those small meteoric masses which an ever-increasing mass of evidence tends to show occupy all the realms of space.' Practically all the theories with regard to the origin of new stars are modifications of one or the other of these ; either an internal convulsion, or an external collision, is hypotheticated.

In 1887 Prof. Lockyer came to the conclusion that 'new stars, whether seen in connection with nebulæ or not, are produced by the clash of meteor-swarms, the bright lines seen being low-temperature lines of elements, the spectra of which are most brilliant at a low stage of heat.' On this theory, the temperature produced by the collision must depend upon the density of the meteor-swarms in action, and will gradually decrease from the time at which the densest parts cross one another. The spectroscopic observations should therefore indicate a reduction of temperature as 'Novæ' diminish in brilliancy. A detailed discussion of the spectra of new stars shows that this decrease of temperature really occurs, and follows the same sequence as would be expected by the cooling of a swarm of meteorites suddenly made hot and then left to cool.

Nova Coronæ exhibited several bright lines in its spectrum when it first appeared in 1866. One by one these faded away, and a fortnight after the first observation it was found that the spectrum was precisely the same as that produced by adding the spectrum of a nebula to that of a comet.

Nova Cygni showed a large number of bright lines in its spectrum. With the exception of one, all the lines faded away as the Nova decreased in brightness. This line increased in intensity, it thrived under conditions which killed its companions. More remarkable still is the fact that the line was coincident in position, and doubtless identical with the chief line of the spectrum of nebulæ. A little more than twelve months after the first observation, only one line remained in the spectrum of Nova Cygni, and that the one which had brightened as the others faded. The new ' star ' had become indistinguishable from a nebula.

Nova Andromedæ appeared like a bright point near the centre of the nebula of Andromeda. Its spectrum was almost the same as that given by the nebula itself, and by many comets. It consisted chiefly of bands due to carbon. A hydrogen line was seen, however, which does not occur in the spectrum of this particular nebula.

The new star which appeared in February, 1892—Nova Aurigæ —exhibited a spectrum of a very remarkable character. Hydrogen was the most conspicuous element, and the lines due to this and other substances were found to have dark lines on their more refrangible sides. This anomalous appearance was seen both in the photographic and the visual spectrum. In order to account for it, various theories have been propounded by spectroscopists and others. A brief survey of these theories was given by one of the authors in *Nature* of May 4, 1893, and is reprinted in the following paragraphs.

Lockyer's and Huggins' Interpretations of the Spectrum of Nova Aurigæ.—The interpretation naturally put upon a composite spectrum like that exhibited by Nova Aurigæ was that two discrete masses were engaged in producing the body's light; one, having a spectrum of dark lines, was rushing towards the earth, while the bright-line star or nebula was running away. As Prof. Lockyer remarked in a paper communicated to the Royal Society on February 7, 1892, 'the spectrum of Nova Aurigæ would suggest that a moderately dense swarm [of meteorites] is now moving towards the earth with a great velocity, and is disturbed by a sparser one which is receding. The great agitations set up in the dense swarm would produce the dark-line spectrum, while the sparser swarm would give the bright lines.' In spite of its simplicity, however, and its ability to account for the observed facts, the meteoritic theory has not commended itself to the minds of some astronomers. Dr. Huggins favours the idea that the outburst was the result of eruptions similar in kind to those upon the sun, but the acquisition of knowledge of the light changes of stars forced him to withdraw the suggestion that the luminosity of a Nova is produced by chemical combustion, in fact, to relinquish entirely the crude conception of a burning world propounded in 1866. In its place Dr. Huggins put the view that Nova Aurigæ owed its birth to the near approach of two gaseous bodies. ' But,' he admits, ' a casual near approach of two bodies of great size would be a greatly less improbable event than an actual collision. The phenomena of the new star scarcely permit us to suppose even a partial collision, though if the bodies were diffused enough, or the approach close

enough, there may have been possibly some mutual interpene-
tration and mingling of the rare gases near their boundaries.'

'An explanation which would better accord with what we know
of the behaviour of the Nova may, perhaps, be found in a view
put forward many years ago by Klinkerfues, and recently
developed by Wilsing, that under such circumstances of near
approach enormous tidal disturbances would be set up, amounting,
it may be, to partial deformation in the case of a gaseous body,
and producing sufficiently great changes of pressure in the
interior of the bodies to give rise to enormous eruptions of the
hotter matter from within, immensely greater but similar in kind
to solar eruptions.'   Serious objections to the Klinkerfues-Wilsing
hypothesis have been pointed out by Herr Seelinger.   He has
shown that the statical theory of tides that has been applied is
entirely inappropriate to the case, and also that the hypothesis
involves assumptions amounting almost to impossibilities.   In the
first place, the pairing of the bright and dark lines makes it
necessary to assume that the two bodies engaged were of similar
chemical constitution, one having an absorption spectrum and
the other an equivalent radiation spectrum.   But even if we
make this unthinkable supposition, a fatal objection has been
pointed out by Mr. Maunder.   It is that the bright lines ought to
have had their refrangibility increased, not decreased as the spectroscopic observations show them to be.   In other words, the
erupted matter would approach the earth, not recede from it.
This single undisputable fact effectually disposes of the chromospheric hypothesis to which reference has been made.

Sidgreaves' Chromospheric Theory.—Another chromospheric theory in which only a single star is involved has been
put forward by Father Sidgreaves.   He says, 'It is only necessary
to consider the conditions under which the blue-side shift of the
Nova's lines should produce the absoption effect while the red-
side parts show unclouded radiation.   A great cyclonic storm of
heated gases would produce this double if the heated gases were
rushing towards us in the lower depths of the atmosphere trending
upwards and returning over the stellar limb.   In the lower
positions the advancing outrush would be screened by a great
depth of absorbing atmosphere, while as a high retreating current
its radiation would be along a clear line to our spectroscopes.'
This explanation is plausible enough, but it does not go to the
root of the matter.   How, for instance, does Father Sidgreaves

account for such a tremendous eruption as that required by his hypothesis? It is difficult to believe that internal forces could sustain, for two months, a stream of gas rushing earthwards with a velocity of about 400 miles per second, and then curving round and receding at the rate of 300 miles per second. And the idea becomes still more incomprehensible when we remember that the body possessing this marvellous store of energy was quite invisible before December, 1891. Until Father Sidgreaves explains the machinery by which the terrific whirl of chromospheric matter was started and kept up, his theory can hardly be seriously discussed.

Modifications of the Meteoritic Theory.—Mr. W. H. Monck has suggested that a star, or a swarm of meteors, rushing through a gaseous nebula, afford the best explanation of the phenomena. The only difference between this idea and that of Prof. Lockyer's is that the nebula is supposed to consist of gaseous instead of meteoritic particles. But, from a dynamical point of view, there is no distinction between the two, for it is well known that Prof. G. H. Darwin has proved that the individual meteorites of a swarm would behave like the individual particles of a gas. Referring to the collision with a gaseous nebula, Mr. Monck says:—'The previous absence of nebular lines, even if clearly proved, would not be conclusive as to the non-existence of such a nebula, for its temperature may not be high enough to produce these lines until raised by the advent of the star. A considerable proportion of Novæ, however, appear to be connected with known nebulæ. Irregularities in the nebulæ would produce the observed fluctuations of light, and if the relative velocity was considerable the bright gas-lines of the nebula would be distinguishable from the dark absorption lines of the star. The bright lines would be broader than usual, because the velocity of the portion of the nebula adjoining the star would be partially destroyed and the luminous gas would thus be moving with different velocities. The heating being confined to the surface of the star, the cooling would take place more rapidly than after an ordinary collision. But if the star travelled far through the nebula in a state of intense incandescence, portions of the surface would from time to time be vaporised and captured by the nebula, the mass of the moving star thus diminishing at every step. It might even end in complete vaporisation, as meteors are sometimes vaporised in our atmosphere. Herr Seelinger has worked out mathematically a theory very similar to that of Mr. Monck. He supposes that a body

E

enters a cosmic cloud, such as have been shown by photography to be widely scattered through space. Whatever the constitution of such a nebulous mass, collision with it causes an increase of temperature, and a vaporisation of some of the constituents of the colliding body.. The process is precisely similar to the entrance of a meteor into the earth's atmosphere. According to Herr Seelinger, Nova Aurigæ was produced in this wise. A dark body was rushing earthwards through space; it came to a mass of nebulosity, the light of which was so feeble that the eye could not appreciate it; the collision caused an increase of temperature and of luminosity; the heaping up of the glowing vapours in front of the colliding body produced the spectrum of dark lines, and the bright-line spectrum was given by the vapours left behind as the body moved onwards. These vapours would quickly assume the velocity of adjacent parts of the nebula, hence the dark lines would appear on the more refrangible sides of the bright ones in the manner observed.

Mr. Maunder also favours a collision theory, his idea being that a long and dense swarm of meteors rushed through the atmosphere of a star, and produced the phenomena exhibited by Nova Aurigæ. As the stream passed periastron, the spectrum of the glowing meteorites, and that of the constituents of the stellar atmosphere with which they were colliding, would appear together with the absorption spectrum of the star.

Objections to the Meteoritic Theory.—From what has been said it will be seen that none of the collision theories are substantially different from that laid down by Prof. Lockyer in 1877. It has been asserted that the meteoritic theory is not competent to explain the observed facts, one of the commonest objections being that the collision of two meteor swarms would be accompanied by a very considerable slackening of the rate of movement. Against this can be urged Seelinger's proof that the great relative velocity indicated by the spectrum could remain practically unchanged, and, in spite of this, enough kinetic energy could be transformed into heat to cause a superficial incandescence. Another objection is that it is impossible to conceive of meteor swarms of such magnitude that though rushing through one another with a relative velocity of more than seven hundred miles per second, disentanglement did not take place until two or three months had elapsed. But the long-exposure photographs taken in recent years show that space is full of nebulous matter, and the 'stream of tendency' is towards the idea that such masses are not gaseous but of meteoritic constitution. Now a simple calculation proves that even if Nova Aurigæ had a parallax of one second of arc, the whole of the luminosity received up to the end of April, 1892, could have been produced by the collision of two bits of nebulous matter each of which would subtend an angle at the earth of less than half a minute of arc. Surely it is not too much to assume the existence of meteoritic swarms of such comparatively small dimensions.

Support of the Meteoritic Theory.—In some incidental remarks upon temporary stars, Mr. Maunder agreed with Prof. Lockyer in 1890 that they 'must be stars in quite another sense to our sun. The rapidity with which their brightness diminishes is plain proof of this. Only small bodies could cool so rapidly, and since despite their vast distance (for their parallax is insensible) these Novas show themselves conspicuous, we are obliged to explain their brilliancy by considering them as consisting of aggregations of such small bodies ; the total extent and mass of the swarm making up for the insignificant size of its components.'

It will be seen that Prof. Lockyer's theory fits in with these observations most aptly. ' New stars,' he says, ' whether seen in connection with nebulæ or not, are produced by the clash of meteor swarms. Clearly, as the swarm cooled down after the collision, we should find its spectrum tend to assume the nebular type.' It is quite immaterial whether the chief nebular line is considered to be due to magnesium or not. According to the meteoritic hypothesis, a new star, as it diminishes in brilliancy, and presumably in temperature, must degrade towards the condition of a nebula. Accept the observations in proof of such a transformation, and the idea that nebulæ are entirely composed of glowing gas becomes untenable, unless it is believed that a Nova increases in temperature as it diminishes in brightness. On the other hand, the change of a new star into a nebula gives strong support to Prof. Lockyer's view that nebulæ are low temperature phenomena. It was therefore expected that Nova Aurigæ should assume the characteristic badge of a nebula. The expectation has been strikingly realised. In August, 1892, the star revived, and on the 19th of that month Prof. Campbell, of the Lick Observatory, wrote the following account of his observations of it :—' The brightest line previously observed was resolved into three lines, whose wave-lengths were about 501,

496, and 486, which were at once recognised to be the three characteristic nebular lines. The same morning Prof. Barnard, using a 36-inch equatorial, observed the Nova as a nebula $3''$ in diameter, with a tenth magnitude star in the centre. Thus the nebulous character of the object was independently established by two entirely different methods.' Writing on the same subject, Prof. Barnard has remarked :—' I think it unquestionable that had any decided nebulosity existed about the star at its first appearance, it would have been detected in observations with the 36-inch, especially when the star had faded somewhat. So it is clearly evident that there has been an actual transformation in every sense of the word of a star into a nebula within an interval of only four months.' Herr Renz also observed the nebular character of the Nova by means of the Pulkowa refractor. On the other hand, one or two observers were unable to detect the nebulosity, and it does not appear on Dr. Robert's photograph of the region. It is impossible, however, to think that an observer of Prof. Barnard's calibre could have been deceived in the matter ; hence the conflicting observations are probably accounted for by fluctuations in the extent and brightness of the nebulosity.

The spectroscopic evidence of the nebular character of Nova Aurigæ in its old age does not rest merely upon Prof. Campbell's observations. Prof. Copeland examined the spectrum on August 25 and 26, and also Mr. J. G. Lohse. From the measures obtained the mean values assigned to the two brightest lines were λ 500·3 and λ 495·3, while a fainter line was seen in the position λ 580·1, which is also the position of a bright line found in the Wolf-Rayet stars and Nova Cygni. Mr. Fowler has also observed the two lines at λ 5006 and λ 4956. But perhaps the most convincing of all testimonies is contained in a paper by Herr Gothard on the spectrum of the new star in Auriga as compared with the spectra of planetary nebulæ. The author photographed the spectra of a number of nebulæ, and compared the results with his photographs of the Nova spectrum. 'Each new photograph,' said he, ' increased the probability, which may be considered as a proved fact, that the *spectrum not only resembles, but that the aspect and the position of the lines show it to be identical with the spectra of the planetary nebulæ*. In other words the new star has changed into a planetary nebula.' In the face of this array of facts nothing could appear to be more satisfactorily established than the descent of the Nova to the condition of a nebula. It must be added, however, that Dr. Huggins has made observations which have led him to believe that 'the bright band in the Nova spectrum is resolved into a long group of lines extending through about fifteen tenth-metres' when a high dispersion is employed. Such are the theories with regard to the origin of Nova Aurigæ and new stars generally. From the survey we see that any and all chromospheric theories fail to explain the transformation of the Nova into a nebula, so they should be abandoned. And finally, the whole sequence of spectroscopic phenomena is explainable on the hypothesis that the light was produced ' by the clash of meteor-swarms.'

Novæ in Relation to Comets and Nebulæ.—We have seen that the spectra of Novæ frequently resemble those of comets and nebulæ. Carbon, hydrogen, and magnesium are the chief elements exhibited by each of these classes of celestial bodies. Not only is this so, but the majority of nebulæ and some comets at aphelion give the same spectrum as that of Nova Cygni when at its lowest temperature. It is argued from this that nebulæ are low-temperature phenomena. And since it is generally accepted that comets are swarms of meteors in our system, it seems probable that nebulæ and new stars, which are often indistinguishable from comets in their spectra, are swarms of meteors out of our system.

A comet in passing from aphelion to perihelion has its spectrum changed, owing to the increase of temperature brought about by the condensation of the meteor-swarm of which it consists. The swarm of meteors which constitutes a nebula goes through similar changes during condensation to a star. In each of these cases, the changes can be matched by gradually increasing the temperature of meteorites in the laboratory and observing their spectra. Two or more swarms are involved in the production of a new star, so a mixed spectrum is obtained depending upon the degree of condensation of each swarm. Thus it is that Novæ give different spectra. On any probable supposition, the highest temperature occurs at maximum brilliancy. The 'star' afterwards sinks back into the condition in which it existed before the disturbance. Its changes should therefore have an opposite sequence to that passed through by the condensing swarms of meteorites which constitute comets and nebulæ. This is found actually to be the case, both as regards changes in spectrum and colour, and goes to show that Prof. Lockyer's meteoritic origin of new stars is a very probable one.

QUESTIONS ON CHAPTER II.

1. Give an account of the recent observations of 'New Stars,' and the conclusions to which they lead. (1893.)

2. State the chief applications which are now made of photography in astronomical observations. (1892.)

3. Give an account of recent researches in connection with the origin of new stars. (1892.)

4. How have stars been classified in relation to their spectra? (1890.)

5. Give an account of recent researches in connection with the distances of the stars. (1890.) (See p. 34.)

6. Describe some recent results obtained by means of long-exposure photographs of the heavenly bodies. (1889.)

7. Give an account of recent researches in connection with the spectra of of stars and nebulæ. (1888.)

8. State what you know about the spectrum of a Lyræ and the sun respectively, and the conclusions which have been drawn from the observations. (1880.)

9. State what you know about the spectrum of Sirius and Uranus respectively, and the conclusions which have been drawn from the observations. (1879.)

10. What spectroscopic changes take place, according to the meteoritic hypothesis, during the condensation of nebulæ?

11. State some of the causes that have been suggested to account for the origin of 'New Stars.'

12. Give an account of the researches which led Lockyer to conclude that the chief nebular line is due to magnesium.

13. Discuss the evidence in favour of the nitrogen and of the magnesium origin of the chief nebular line.

14. Give an account of recent researches in connection with the photographic spectra of the brighter stars.

15. Give an account of recent researches in connection with the motions of stars in the line of sight.

16. State the principle of the method employed in the determination of stellar motions in the line of sight.

17. How has it been proved that binary stars exist in the heavens so close together that no telescope can show them to be double?

18. What are spectroscopic double stars, and how have they been discovered?

19. Give an account of recent researches in connection with the motion of Algol in the line of sight.

20. What constitutes the system of Algol and how have the different members been discovered?

21. Give an account of Pickering's conclusions with regard to the composite character of the spectra of certain stars.

22. Give a general account of the spectra of new stars that have been observed.

23. State briefly some of the causes that have been suggested to account for the origin of Nova Aurigæ.

24. Describe the relations between new stars, comets, and nebulæ.

CHAPTER III.

CONCERNING MEMBERS OF THE SOLAR SYSTEM.

THE notes in this chapter relate to the sun, moon, planets, and comets, and supplement several chapters in *Advanced Physiography*. Much of the matter herein contained, however, has not yet found a place in any text-book of Physiography, though it has formed the subject of numerous questions in the Honours stage.

Janssen's Solar Photographs.—Some of the most interesting results on the structure of the sun's surface have been obtained photographically by Dr. Janssen at Meudon, near Paris. Dr. Janssen's method of taking solar photographs differs somewhat from that employed by other astronomers. He utilises the fact that an ordinary photographic plate is most sensitive to rays near the violet end of the solar spectrum. Only the rays which produce the maximum chemical effect are allowed to act upon the plate. This practically means that light of one colour is used, for the most actinic rays only extend over a small region. Hence it is not necessary to have an absolutely achromatic object-glass. The telescope used by Dr. Janssen has an aperture of only five inches, but by means of an enlarging lens, marvellous pictures of the sun's disc a half a yard in diameter are obtained by exposing wet collodion plates from $\frac{1}{200}$ to $\frac{1}{1000}$ of a second. The great advantage which such pictures possess over the best visual observations is that they exhibit the whole of the sun's visible hemisphere at one view, and so bring out relations which are overlooked when only a limited part of the surface is observed at once.

An examination of Dr. Janssen's pictures shows that a large portion of the solar surface has a hazy appearance (Fig. 9). This undefined appearance, or 'smudginess,' as it has been termed, was at first considered to be due to peculiarities in the wet collodion film, or to currents of air in the telescope tube, or vapour rising from the damp collodion when the sun's rays fell upon it. To settle the question, Dr. Janssen has taken a number of pictures in succession, and he finds that the haziness occurs in the same

regions on all the series. Clearly this could not be the case if
the effect were produced by accidental disturbances of the air in
the telescope, or deficiencies in the uniformity of the collodion
film. But though the 'smudges' occupy practically the same
regions and present almost the same appearances on pictures
taken a few minutes after one another, they change far more
rapidly than sun-spots or faculæ. Photographs taken with an
interval of an hour or so exhibit striking differences in the

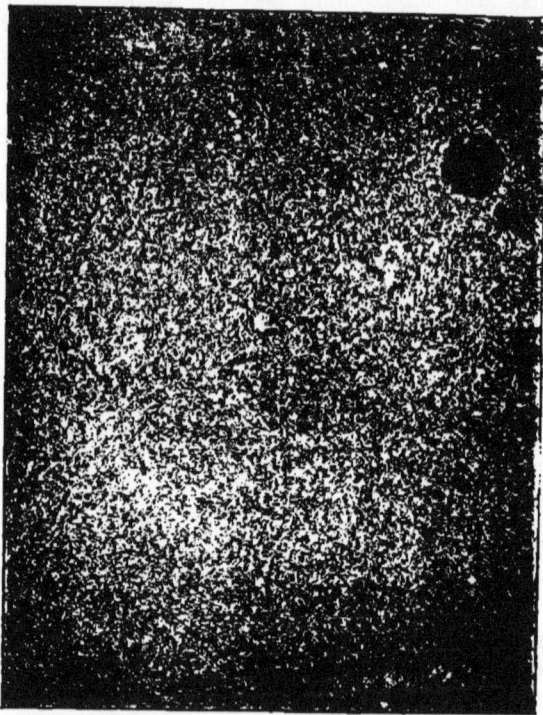

Fig. 9. A view of a Sun-spot and a part of the Solar photosphere, from a
photograph by Dr. Janssen, reproduced in *Knowledge*.

character of the hazy parts. The smudginess seems to be in a
state of continual change, and the variations are especially well-
marked in the neighbourhood of spots and faculæ.

Dr. Janssen believes that the indistinctness is produced by the
trembling or shifting of a portion of the sun's atmosphere. The
parts of the surface which have a well-defined appearance upon

the pictures are probably not overlaid by agitated regions. Certain it is that hazy areas perpetually shiver and shift over the sun's surface, and the discovery of the fact is one of the triumphs of solar photography. As to the cause of the phenomenon Prof. Young gives the following opinion :—

'It is not, however, certain that the disturbed portions of the solar atmosphere, which produce the indistinctness in question, lie near the sun's surface. It may be that they are high up, and it would not be an unreasonable conjecture to suppose that the streamers and luminous masses of the corona may be concerned in the phenomenon ; it is almost certain that any great aggregation of chromospheric matter would modify the appearance of whatever might be situated beneath it. The simple fact is, of course, that we are looking down upon the granules and other features of the sun's surface, not through an atmosphere shallow, cool, and quiet like the earth's, but through an envelope of matter, partly gaseous and partly, perhaps, pulverulent and smoke-like, many thousand miles in depth, and always most profoundly and violently agitated.'*

Solar Prominence Photography.—During the last two or three years remarkable advances have been made in the photography of solar prominences in ordinary daylight. Professor George E. Hale, of Kenwood Observatory, Chicago, has obtained results which surpass all others (Fig. 10). In order to understand the principle of his method, it must first be remembered that the K line is always reversed, that is, bright, in the spectrum of a prominence. It is this fact that has rendered the photography of prominences possible. The 'spectroheliograph' devised and used by Professor Hale consists essentially of two movable slits, one at the focus of the collimator of a powerful grating spectroscope, and the other just within the focus of the view telescope. The whole instrument is rigidly attached to the lower end of a telescope, forming a kind of tele-spectroscope. To photograph the prominences round the sun at any instant with Professor Hale's arrangement, the centre of the sun's image formed by the equatorial telescope is made to coincide with the axis of the collimator and is kept in this position by the driving clock. The light from the sun's visible disc is then shut off by means of a circular diaphragm, and a clepsydra causes the collimator slit to move gradually across

* *The Sun*, p. 112.

the image of the sun, while the second slit, near the focus of the view telescope, moves at such a rate that the light of the K line constantly falls upon a fixed photographic plate which replaces the eye. One slit thus passes across the sun's image in search of prominences round the limb, while the function of the other is to prevent any light but K light from falling upon the photographic plate behind it. In this way the forms of prominences are built up,

Artificial Eclipse.   The Eclipse of 1860.

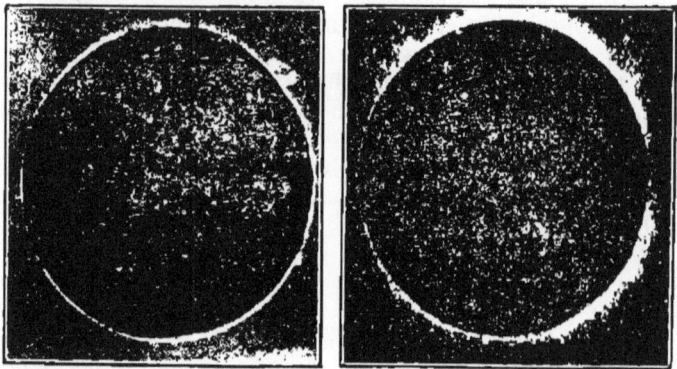

Fig. 10. Comparison of a photograph of the chromosphere, obtained by Prof. Hale in full daylight, and one taken during the total eclipse of 1860.

as it were, by successive stages, and a single photograph shows all the luminous projections from the sun's limb at the time of exposure. Professor Hale has extended his method still further. He has found that the H and K lines are always bright in the spectra of faculæ—those shreds and patches of brilliant matter which are usually well visible near sun-spots and the sun's edge—and has used the fact to obtain photographs of them irrespective of their position. This end is attained in the following manner. After a photograph of the prominences and chromosphere has been taken by the operations described above, the diaphragm is removed and the slits are again set in motion, but at a much higher speed. Images of all regions on the sun in which the K line is reversed thus act upon the photographic plate, and, upon development, a permanent and trustworthy record of the state of the sun at the time of exposure is obtained. By the adoption of this method of research, Professor Hale has discovered faculæ quite invisible to the eye which frequently burst out upon the sun and float over sun-spots. He also finds that faculæ are much more common solar phenomena that has been supposed. They are far more extensive than spots and considerably exceed spots in point of size.

Changes observed during a Sun-Spot Period.—Sun-spots exhibit the most obvious changes during the 11-year cycle of solar activity. At a minimum epoch, the proportion of spotted to unspotted surface is small; and several weeks may pass without a spot being seen. A gradual increase of spotted area then sets in and continues for about four years, when a maximum is reached, and scarcely a day passes without the appearance of a number of spots. From the epoch of maximum activity, there is a gradual decrease in the frequency and extent of spots. This continues for about seven years, when another minimum is reached. The interval between two successive maxima or minima is thus about eleven years. After a minimum, sun-spots occur most frequently in the parts of the spot zones most removed from the solar equator, that is, near 40° of north and south solar latitude. The zones of greatest frequency then suffer a decrease of latitude as the period goes on, reaching their lowest limit at the epoch of the following minimum, after which spots again break out in high latitudes. The character of spots also appears to undergo a periodic change, for spots at minimum are generally more or less symmetrical in shape, whereas those on the sun near a maximum are broken up as if they were the centres of a more violent action. The photosphere is also affected. Near a minimum, the photospheric clouds have a 'willow leaf' structure, whereas at a maximum a 'rice grain' appearance is seen. Faculæ and eruptive prominences undergo similar variations to those exhibited by spots, both as regards extent and distribution in latitude. Veiled spots and quiescent prominences do not appear to be affected by the rise and fall of solar activity.

The Distribution of Solar Spots and Prominences.— Sun-spots are practically confined to two zones between 5° and 45° north and south of the solar equator. Eruptive or metallic prominences are also very rarely found outside these limits. But for quiescent or 'quiet' prominences the case is different. These appear to favour no particular latitude, and are equally distributed from the sun's equator to the poles. The distribution of spots and metallic prominences varies throughout a cycle of solar activity. After a minimum, both occur most frequently near the parts of the zones most removed from the equator. There is then a gradual descent of the latitudes in which the phenomena are most frequent until the following minimum epoch is reached, when spots and metallic prominences are often seen near the equator. At the same time spots appear in high latitudes, a new cycle beginning before the old has died out. The connection between spots and metallic prominences is also brought out by observations of single years. In some years the spots are more frequent in the northern than in the southern hemisphere of the sun, or *vice versâ*. At other times spots occur with almost equal frequency in both hemispheres. But in all cases the distribution of metallic prominences follows that of spots. If the maximum of spots occurs in the north hemisphere, the maximum of prominences is also found there ; if in the south hemisphere, the prominences change their maximum latitude in the same manner. Quiet prominences, on the other hand, are affected in no way by the changes in the position of the zone of maximum sun-spot frequency, either during a solar cycle, or during the year ; and are thereby distinguished from prominences of the eruptive kind.*

Prof. Spoerer has made a comprehensive investigation of the relation between solar spots and prominences. He has shown the results of his research by means of the curves of Fig. 11. The rise and fall of the upper curve exhibits the increase and decrease respectively of solar spottedness during the years indicated on the horizontal line. The two lower curves show the decline of prominences towards the equator in passing from one minimum to the next.

Lockyer's Observations of Sun-Spot Spectra.—Two different methods are employed in recording sun-spot spectra. Every line widened in the spectrum of a spot may be observed and tabulated, or a fixed number only of the lines most affected may be picked out. The latter method is employed by Professor Lockyer. Two regions of the solar spectrum are chosen, one between Fraunhofer's F and b, the other extending from b to D. The six *most widened* lines in each of these regions are discrim-

---

*See *Chemistry of the Sun.* Lockyer. pp. 419-422.

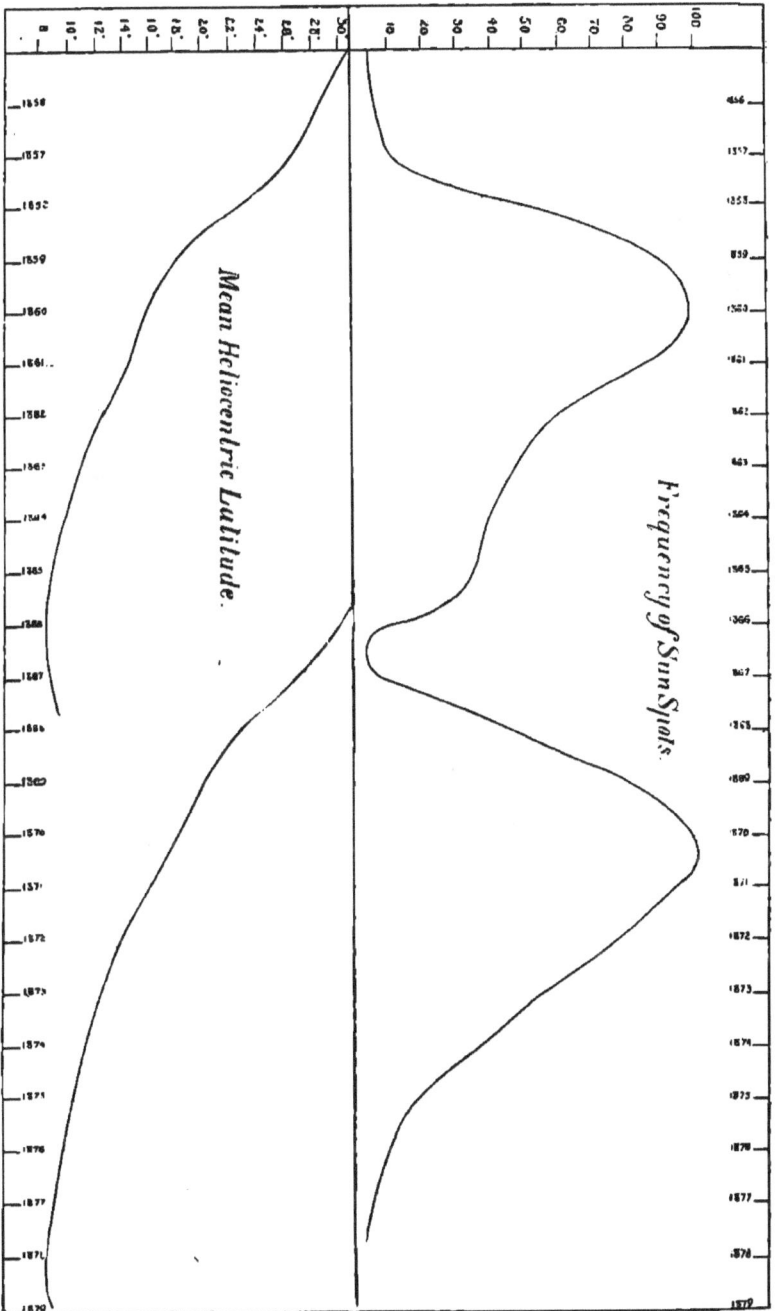

Fig. 11. Spoerer's curves showing the relation between the frequency and the mean latitude of Sun-spots.

inated in the case of every spot large enough to be observed. This is done without any reference being made to the origins of the lines affected, that is to say, the observer does not attempt to find whether the six most widened lines are due to iron, or nickel, or any other element. A discussion of some hundreds of observations made in this manner, and extending over nearly a complete cycle of solar activity has shown that near a minimum, when the sun is coolest and least disturbed, the most widened lines belong to elements such as iron, nickel, and titanium, with which we are familiar on the earth. In passing to a maximum epoch, however, the most widened lines are not representatives of terrestrial elements, but are lines the origins of which are unknown. This change will be better understood from an example. In 1879 the sun was in a minimum state of activity, and in the total number of spots observed during that time, sixty different lines representing iron were found to be among the most widened. In every spot, lines of iron were seen to be most affected. About the time of maximum in 1885, the case was very different. Only four spots out of one hundred observed had spectra in which iron lines were the most widened, and only three different iron lines, instead of sixty, were so distinguished. Nickel, titanium, and other elements known to us disappeared from the spectra in a similar manner. Professor Lockyer ascribes these variations to dissociation. He believes that iron and nickel and titanium, and many other substances known to us as elements, are really complex bodies. At the time of minimum activity the vapours of these bodies can exist as such. But during a maximum, as these vapours descend to the photosphere to form a spot, they are decomposed or dissociated, and only those constituents which can exist at a very high temperature remain, and reveal their presence as lines of unknown origin in the spot spectrum.*

The Rotation Periods of Mercury and Venus.—Nearly a century ago Schröter published his observations of the physical aspect of Mercury and assigned to the planet a period of rotation. In 1889, Schiaparelli an Italian observer, communicated to the world the results of his observations of the planet since 1882. As many as 150 drawings were made of certain markings on Mercury. The markings appeared to be identical in aspect when

---

* Professor Lockyer's communication to the Royal Society on this subject is reprinted in *Nature*, vol. xxxiv., 1886.

observed at the same hour on consecutive days. To account for this, three hypotheses have been propounded.

(1.) That the time of rotation is about 24 hours.

(2.) That the planet makes two or more rotations in the same interval.

(3.) That the time of rotation is so slow as to be inappreciable when observing the markings during a few days.

Schröter decided in favour of the first hypothesis, and Bessel, from a discussion of this observer's data, determined the time of rotation to be 24h. om. 52·97s. Schiaparelli favours the third hypothesis. From an examination of the drawings by others and himself he concludes that Mercury revolves round the sun in the same manner that the moon revolves round the earth, always presenting to it the same side; hence, since the planet's periodic time is 87·9693 days, this must be the time of rotation on its axis.

On account of the eccentricity of the moon's orbit, we get lunar libration in longitude. In a similar manner, the large eccentricity of the orbit of Mercury causes the planet to have a large libration with respect to the sun. Viewed from the sun, about 23½ degrees are seen round the east and west edges of the planet, which would be invisible if no libration existed. On Schiaparelli's hypothesis, the whole of the surface is never illuminated, but only that part always facing the sun, and 23½° on either side of it.

Schiaparelli has also made an extended enquiry into the question of the rotation of Venus. His investigation and his own observations have led to the conclusion that the time of rotation of the planet is 224·7 days, that is to say, Venus, like the moon and Mercury, rotates on her axis in the same time that she takes to make a sidereal revolution; the axis of rotation being nearly perpendicular to the plane of the orbit. From an examination of the observations of previous astronomers, Schiaparelli found that those records which have been supposed to fix the rotation period as about 24 hours are open to question. Cassini's observations of bright markings in 1866-67 were shown by the Italian astronomer to have been wrongly interpreted, a discussion of them indicating that they also support a period of rotation of 224·7 days.

The conclusions arrived at by Schiaparelli have not been accepted by all astronomers. More observations are required before the question of the period of rotation of these two planets

can be settled either one way or the other. A difficulty arises
from the fact that Mercury and Venus are so brilliant that the
spots and markings upon them can only be recognised after
careful observation.

The Planet Venus.—Some important observations of the
planet Venus have recently been made by M. Trouvelot and are
detailed in *Nature* of September 15, 1892.

The surface markings of the planet are classified into (1) Snow-
caps at the poles of the planet; (2) whitish spots of a more or
less rounded form; (3) large greyish spots or bands. The
snow-caps resemble those which are a feature of the surface of
Mars. According to M. Trouvelot, 'They have the form of a
uniform white segment of a circle, which, when seen edgeways,
appear as simple lines; they are always exactly 180° apart;
sometimes only one is seen because the other is not lighted up by
the sun; they are of a permanent nature, their disappearances
being due not to their annihilation, but simply to the fact that
they cannot be seen when receiving no light upon them. One
main feature in which they differ from the spots on Mars is that
they neither increase nor decrease with the seasons, at any rate
to a sufficient extent to be sensibly noticed.'

A large grey spot of well-defined outline was described in 1876,
and M. Trouvelot observed one that suited the description in 1891.
He was thus led to believe that the two spots were identical, and
represent a permanent marking upon the planet, frequently
obscured, however, by dense layers of atmosphere. Observations
of this spot indicate that the time of rotation of Venus is rather
more than twenty-four hours, whereas Schiaparelli has assigned
the planet a rotation period of 225 days. It seems very probable,
therefore, that though the rotation and revolution periods of
Mercury may be identical, the identity does not hold good for
Venus. In order to determine whether any irregularities exist on
the surface of Venus, in other words, whether there are
mountains and valleys on the earth's twin-sister, M. Trouvelot
examined the terminator from time to time. If the surface of the
planet were perfectly spherical, the terminator would always be
an unbroken curve. Observations show that this is not the case.
The terminator has frequently been seen notched and wavy, thus
indicating that all parts of the surface have not the same level.
The white spots referred to above appear to be higher than the
surrounding regions, and the greyish spots appear to be lower

than parts near them.   In fact, the former represent mountains of Venus and the latter valleys.

Mars in 1892.—The doubling or 'gemination' of the canals of Mars, first observed by Schiaparelli, was confirmed during the opposition of 1892. Until this confirmation, astronomers have looked askance at Schiaparelli's double canals. It was insinuated that the phenomenon was not real, but due to instrumental deficiencies, or fatigue of the eye, and it was said that if a dark line were drawn upon a piece of white card and looked at steadily, after a short time it would appear double. Since, however, one or two other observers have declared the gemination to be real, several explanations have been suggested to account for it. Professor Lockyer's idea is that the appearance is produced by a belt of cloud lying along the centre of each canal at the time of the apparent doubling. The theory is a very plausible one, because the doubling seems to be connected with the planet's seasons. On this account cloud-belts would sometimes lie along the canal and at other times they would be absent. Mr. Lebour* has pointed out that when glass is broken by torsion, the cracks produced are similar to the channels on Mars. Such fractures are a necessary consequence of the cooling of a heated interior and the shrinking of the planet's crust. 'Mere fractures, such as we meet with in our own planet, could, of course, not be seen from any considerable distance; and if the circumstances of denudation were the same in Mars as with us, the "canals" could certainly not be the representatives of our usually hidden and featureless earth-cracks. There seems, however, to exist in the extraordinarily rapid melting of gigantic ice-fields, described by Prof. Norman Lockyer, some evidence of denuding power in Mars on a scale enormously larger than is the case with us. Earth fractures—and, for the matter of that, Mars fractures, too—must many of them be lines of weakness along which denudation acts more freely than elsewhere; and if this denudation be phenomenal and cataclysmic, as appears to be likely in Mars, wide valleys or channels capable of being distinguished at great distances would soon be scoured out along them.' Another possible cause has been suggested by M. Stanislas Meunier. He has taken a metallic sphere and traced upon it lines and spots similar to those on Mars, but not doubled. Surrounding the sphere, at a distance of about a tenth of an inch from its

---

* *Nature*, October 27th, 1892.

F

surface, a fine transparent layer of muslin was fixed. This was supposed to represent the planet's atmosphere. When the sphere was looked at through the muslin, all the lines and spots were seen double, and by undulating the muslin several peculiarities were noticed which have been observed by Schiaparelli on the planet Mars. M. Meunier's explanation is that 'the solar light is reflected from the planet's surface very unequally, that from the continents exceeding that emitted by the deeper parts, seas and canals. Although the atmosphere is a limpid one, we are unable to see its motions; but if the aerial envelope includes a transparent veil of fog at a suitable height, a contrast would be produced, as was the case with the muslin, by the production of shadows which, to an observer not in a line with the rays reflected from each of the surfaces of small reflecting power, would appear as parallel images.'

The observations of Mars made by Prof. W. H. Pickering at the Arequipa mountain observatory, led him to the following conclusions:—

(1.) The polar caps are clearly distinct in appearance from the cloud formations, and are not to be confounded with them. (2.) Clouds undoubtedly exist upon the planet, differing, however, in some respects from those upon the earth, chiefly as regards their density and whiteness. (3.) There are two permanently dark regions upon the planet, which under favourable circumstances appear blue, and are presumably due to water. (4.) Certain other portions of the surface of the planet are undoubtedly subject to gradual changes of colour, not to be explained by clouds. (5.) Excepting the two very dark regions referred to above, all of the shaded regions upon the planet have at times a greenish tint. At other times they appear absolutely colourless. Clearly marked green regions are sometimes seen near the poles. (6.) Numerous so-called canals exist upon the planet, substantially as drawn by Prof. Schiaparelli. Some of them are only a few miles in breadth. No striking instances of duplication have been seen at this opposition. (7.) Through the shaded regions run certain curved branching dark lines. These are too wide for rivers, but may indicate their courses. (8.) Scattered over the surface of the planet, chiefly on the side opposite to the two seas, a large number of minute black points have been found. They occur almost without exception at the junctions of the canals with one another and with the shaded portions of the planet. They range from thirty to one hundred miles in diameter, and in some cases are smaller than the canals in which they are situated. Over forty of them have been discovered, and for convenience they have been termed lakes.

Little beyond the facts enumerated by Prof. Pickering is at present known about Mars. The white region round the south pole —usually supposed to be a cap of ice or snow—was measured by several observers during the opposition of 1892, and found to decrease in size very rapidly while the planet was passing from its spring equinox to the summer solstice of its southern hemisphere. A number of bright spots, similar to the polar caps, were discovered by M. Perrotin, and seen by him and other observers projecting from the planet's edge to heights between twenty and forty miles. Extensive changes seem to have taken place in various regions of the planet, but this may possibly be due to the shifting of banks of cloud.   On the whole, it seems to be established that many of the chief markings are permanent, and also that real changes take place in other markings which cannot be explained by interpreting the yellowish region as land, and the darker greenish patches as seas.   Prof. Schaeberle has suggested that the reverse is the case, but his idea has received little support.

Comets and the Changes they undergo.—Comets are swarms of meteorites captured from outer space by one of the major planets and revolving round the sun in parabolic or elliptic orbits.   In space they have no luminosity, but the tidal action set up after entrance into the solar system causes the individual meteorites to graze against each other.   Heat is developed by the contact, some of the constituents of the meteorites are volatilised and rendered luminous, and the previously invisible swarm of meteorites becomes a visible comet.

The changes undergone by a comet during its journey round the sun are (1) changes in the appearance of the head and tail, (2) changes in the spectrum.   We will describe these in order.

When a comet is first observed telescopically it looks like a patch of luminous haze, and cannot be distinguished from a globular nebula.   The first change as the comet moves towards the sun is the development of a brighter part, termed the nucleus, near the centre of the patch of nebulosity.   Jets of luminous matter are then shot out towards the sun from the nucleus, and more or less concentric layers of similar material appear to surround the nucleus on the sunward side.   These bend back to form the tail, behaving as if the sun forcibly repelled the material of which they

are composed. The action increases in violence as the sun is
approached, and, as a rule, the tail increases in size. Sometimes
several tails are seen, all pointing away from the sun, but
differently curved. There is reason to believe that the tail
driven straight back from the sun consists of hydrogen. In
the plume-like tails, carbon in some form or other is the
chief constituent. Short, stubby tails consist of dense vapours,
such as iron-vapour. The increase of luminosity goes on
until a few days after the comet has passed its perihelion point.
There is then a gradual diminution in the intensity of the phenomena and the size of the comet, and, when aphelion is again
reached, the object will have sunk to the insignificant patch of
nebulosity from which it grew.

Next as to spectroscopic changes. The general spectrum of a
comet consists of three bright bands due to carbon. This shows that
a comet's light is not merely reflected sunlight. Professor Lockyer
has brought together and discussed all the cometary spectra ever
observed. Each comet has a particular time at which it passes its
perihelion point, that is, the point in its orbit nearest the sun.
Now there can be no doubt that a comet is hotter when near the
sun than when far away from it. A month before perihelion passage the temperature would be lower than a week before the
passage. By arranging the spectra of comets according to the
date of observation with respect to that of perihelion passage,
Professor Lockyer was able to distinguish differences due to
differences of temperature. A couple of comets at aphelion have
shown a spectrum consisting of a single line, and that line the one
which characterises nebulæ. From this and from the similarity in
the appearance of comets and nebulæ, it would seem that a nebula
is simply a comet—a swarm of meteorites—existing in interstellar
space. As comets approach the sun, they become hotter, owing
to the increased number of meteoritic collisions, and the spectrum
changes. The sequence of changes is precisely the same as that
which takes place during the condensation of a nebula into a star,
and is summed up by Professor Lockyer as follows :—

(1.) Radiation of low temperature magnesium.
(2.) Radiation of carbon and manganese flutings.
(3.) Absorption of manganese and lead flutings.
(4.) Radiation and absorption of lines.

The spectrum of a swarm of meteorites gradually increasing in
temperature would go through the same changes as those exhibited
in the spectra of comets and nebulæ. This goes to prove that

both are swarms of meteorites differently situated. The head of a comet is usually considered to be the densest part of the swarm. The tail consists of the vapours driven off by the heat developed by collisions.

Tidal Evolution.—Prof. G. H. Darwin has shown mathematically that at one time the earth and moon formed a single molten mass, rotating in a period of from two to four hours, and revolving round the sun in a period not much shorter than our present year. Owing to rotational instability, or tidal action set up by the sun, the fluid mass separated into two parts, the smaller to form the moon and the larger the earth. The two bodies were then in contact, and each exerted a tidal action upon the other; while the sun raised tides in both. At that time the moon's period of revolution was slightly longer than the earth's period of rotation, in other words, the month was slightly longer than the day. Tracing the history from that time, Prof. Darwin remarks: 'The axial rotation of the moon is retarded by the attraction of the earth on the tides raised in the moon, and this retardation takes place at a far greater rate than the similar retardation of the earth's rotation. As soon as the moon rotates round her axis with twice the angular velocity with which she revolves in her orbit, the position of her axis of rotation (parallel with the earth's axis) becomes dynamically unstable. The obliquity of the lunar equator to the plane of the orbit increases, attains a maximum, and then diminishes. Meanwhile the lunar axial rotation is being reduced towards identity with the orbital motion.

'Finally her equator is nearly coincident with the plane of her orbit, and the attraction of the earth on a tide which degenerates into a permanent ellipticity of the lunar equator causes her always to show the same face to the earth.'

The course of changes brought about by these mutual actions from the birth of the moon until now is tabulated below, and summed up as follows :—(1) The length of the day has increased from 5·6 hours to 23·93 hours. (2) The moon's period of revolution has increased fron 0·23 days to 27·32 days. (3) The number of days in a month increased from 1 to a maximum of 28·83, and then decreased to 27·40 at the present time. (4) The moon's distance has increased from 1·5 the radius of the earth to 60·4 times the radius.

Number of millions of years ago	Length of sidereal day, in mean solar hours	Length of moon's sidereal period, in mean solar days	Number of days in a month	Distance of moon, in earth's mean radii
... ...	5·60	0·23	1·00	1·5
56·81	6·75	1·58	5·62	9·0
56·80	7·83	3·59	11·01	15·6
56·60	9·92	8·17	19·77	27·0
46·30	15·50	18·62	28·83	46·8
Present time.	23·93	27·32	27·40	60·4

The number of days in a month must go on decreasing until a
month of one day is reached. This great day will be about equal
to fifty-seven of our days. 'After the moon's orbital motion has
been reduced to identity with that of the earth's rotation, solar
tidal friction will further reduce the earth's angular velocity, the
tidal reaction on the moon will be reversed, and the moon's
orbital velocity will increase, and her distance from the earth will
diminish.* We shall then have a condition of things such as
obtains to Mars, in which a satellite revolves round the primary in
less time than the planet itself takes to perform a rotation.

Reconciliation of the Nebular and Meteoritic Hypotheses.—Laplace supposed that the sun once existed as a rotating
mass of incandescent gas. As this solar nebula cooled, it con-
tracted and its velocity of rotation increased, so that in time the
centrifugal force at the equator became equal to the gavitational
attraction and a ring of gaseous material was left behind. The
rings thus left behind eventually condensed to form more or less
globular masses (the planets) which in turn shed similar rings to
form satellites.

This hypothesis starts with the existence of a highly heated gas,
forming a nebula. In 1887, Prof. Lockyer brought a large number
of spectroscopic facts to show that nebulæ were not masses of gas
but swarms of meteorites. Collisions and grazings of individual
meteorites would develop sufficient heat to vaporise some of
the constituents and render them luminous.

Prof. G. H. Darwin has investigated the mechanical conditions
of a swarm of meteorites for the purpose of determining whether
Laplace's and Lockyer's theories as to the origin of worlds could

---

* An excellent summary of Prof. Darwin's researches was given by General
Tennant at the anniversary meeting of the Royal Astronomical Society in 1892,
and is printed in the *Monthly Notices* of the Society.

be reconciled.   The former assumed that the solar nebula was gaseous, the latter showed that, in all probability, it was a swarm of solid particles.   Now whatever the nebula is assumed to be, it must behave like a rotating mass of fluid, that is, like a liquid or gas.   From this it would appear at once that the meteoritic origin is put out of court.   Prof. Darwin's researches have shown that this is not the case.   He finds that a condensing swarm of meteorites would behave in precisely the same manner as a cooling mass of gas, hence the only difference between the 'nebular' hypothesis and the 'meteoritic' hypothesis lies in the constitution of the primitive nebula.   In his words, 'According to the kinetic theory of gases, fluid pressure is the average result of the impacts of molecules.   If we imagine the molecules magnified until of the size of meteorites, their impacts will still, on a coarser scale, give a quasi-fluid pressure.   I suggest then, that the fluid pressure essential to the nebular hypothesis is in fact the resultant of countless impacts of meteorites.'*

The Atmospheres of Planets.—*Mercury* has been shown spectroscopically to have an atmosphere in which water-vapour exists, but it cannot be very extensive or dense.

Venus has an atmosphere very similar to that which surrounds the earth, but denser.   This is proved by the fact that a thin ring of light has been seen to encircle the planet just previous to a transit. The phenomenon is caused by the refraction of sunlight in the planet's atmosphere.   Near inferior conjunction also, the crescent-shaped figure is often seen completed by a thin circular line of light due to atmospheric refraction.   Water-vapour is probably present in the atmosphere of Venus.

The Earth has an atmosphere which is sufficiently dense to produce the phenomena of dawn and twilight at a height of 46 miles, and to render meteorites luminous at a height of 200 miles. The pressure of the atmosphere at sea-level is 15 lbs. per square inch of surface, and diminishes as we ascend.   The chief constituents of the atmosphere are nitrogen, oxygen, carbon dioxide, and water-vapour.   (See Chapter IV.)

Mars has an atmosphere containing water-vapour, but probably not so dense as that of the earth.   The existence of an atmosphere is indicated by a slightly undefined appearance of the planet's terminator.   In the case of a body like the moon, devoid of an

---

* *Nature*, vol. xxxix, pp. 81 and 105, Nov., 1888.

atmosphere, the terminator is seen perfectly distinct. Cloud-like markings have also been seen to obscure for a time the markings on the face of Mars.

Jupiter has a very extended atmosphere. Indeed, from the fact that the planet's mean density is only one-third greater than that of water, it seems probable that there is no sharp line of demarcation between the atmosphere and surface, as there is on the earth. The view is also held that the planet is made up of a small solid nucleus and a very thick shell of atmosphere surrounding it. The 'belts' of Jupiter and other markings are in the planet's atmosphere and not upon its surface. In the yellow and red parts of the spectrum of Jupiter, dark bands are seen, in addition to the lines due to reflected sunlight. The origins of these bands are unknown.

Saturn shows 'cloud-belts' similar to those of Jupiter. Its spectrum is also similar—a prominent feature being a dark band in the red, near the position of a low-temperature band due to iron. The atmosphere is probably very extended for the same reasons as those described in the case of Jupiter.

Uranus appears to have a dense atmosphere, and from the fact that the mean density of the planet is very low, the atmosphere is probably very extensive. Enigmatical dark bands are seen in the spectrum of Uranus and indicate that the planet does not shine entirely by reflected sunlight.

Neptune has probably an atmosphere like that of Uranus, but very little is known on this point. The spectra of the two planets are similar.

Determination of the Moon's Mass by Luni-Solar Precession.—On account of the moon's differential attraction upon the earth's equatorial protuberance, the plane of the terrestrial equator tends to coincide with the plane of the lunar orbit. The sun's differential attraction tends to pull the equatorial plane into the plane of the ecliptic. The result of these two actions is to give to the plane of the equator a retrograde motion round the ecliptic plane. A complete revolution of the line of intersection of the two planes (the equinoctial line) takes place in about 26,000 years, the rate being $50''\cdot38$ per annum. This effect is known as *luni-solar* precession. It is greatest when the disturbing bodies are at the greatest distance from the equator, and nil when they are on the equator. Twice a year then, at the equinoxes, the sun's precessional action is nothing, and twice a year, at the solstices, it has its greatest value. · Similarly the moon's action is zero twice a month, when our satellite is on the equator, and attains a maximum value when at its greatest distance above or below the equator. By comparing the effects produced when each body acts alone, the moon is found to have a disturbing action about 2·2 times as great as the sun. Now it can be shown that a differential disturbing force such as that which produces tides and precession varies directly as the mass of the disturbing body and inversely as the *cube* of the distance.

Hence we have the following equations :—

$$\text{Moon's action} = \frac{\text{Mass of Moon}}{(\text{Moon's distance})^3}$$

$$\text{Sun's action} = \frac{\text{Mass of Sun}}{(\text{Sun's distance})^3}$$

But the moon's action is found to be about 2·2 times greater than that of the sun. And the sun's distance is about 390 times greater than the distance of the moon. The above equations therefore become as follows :—

$$2\cdot2 = \frac{\text{Mass of Moon}}{1^3}$$

$$1 = \frac{\text{Mass of Sun}}{390^3}$$

Hence

$$\text{Mass of Moon} : \text{Mass of Sun} :: 2\cdot2 : 390^3$$
$$= \tfrac{2\cdot2}{59319000}$$
$$= \tfrac{1}{27000000}$$

Therefore the moon's mass is $\tfrac{1}{27000000}$ in terms of the sun's mass.

But the sun has a mass 330,000 times greater than the mass of the earth.

Hence, the mass of the moon in terms of the mass of the earth is $\tfrac{1}{27000000} \div 330,000$ that is $\tfrac{1}{81}$.

QUESTIONS ON CHAPTER III.

1. Give an account of recent researches in connection with the surface conditions of the planet Mars. (1893.)

2. Give an account of Prof. George Darwin's researches on the effects produced on the motions of the earth and moon by tidal action. (1892.)

3. State the telescopic and spectroscopic changes observed in comets as they approach perihelion, and the causes which produce them. (1891.)

4. Give an account of the changes brought about in the right ascension, declination, and longitude of stars by luni-solar precession. (1891.) See *Advanced Physiography*, p. 136.

5. Give an account of recent researches in connection with the rotation of Venus and Mercury. (1891.)

6. What is a comet, and what are the changes produced in it during its journey round the sun? (1890.)

7. How has the mass of the moon been determined from luni-solar precession? (1889.)

8. Give an account of recent researches in connection with Laplace's Hypothesis of the origin of the sun and planets. (1889.) See p. 70.

9. Give an account of recent researches in connection with the spectra of sun-spots. (1887.) See p. 60.

10. State what you know concerning the distribution of spots and prominences on the sun's surface. (1885.)

11. Compare the atmospheres of the major planets. (1886.)

12. Give an account of recent researches in connection with the structure of the solar surface. (1885.)

13. State what you know about sun-spots, and explain the terms, corona, chromosphere, photosphere, faculæ. (1883.)

14. State what you know about the surface of the planet Mars. (1882.)

15. What is known regarding the chemical nature of comets? (1882.)

16. How has the connection between comets and luminous meteors been established, and what is the nature of the connection? (1879.)

17. Describe the spectra of sun-spots and prominences.

18. Give an account of the changes observed in solar phenomena during a sun-spot cycle.

19. Describe the changes that take place in the spectra and distribution of sun-spots in passing from a minimum to a maximum of a solar cycle.

20. Describe the method employed by Hale in order to obtain photographs of solar prominences in ordinary daylight.

21. State some of the results obtained by means of the spectroheliograph.

22. Give an account of recent researches in connection with the surface appearance of the planet Venus.

23. State the relations that exist between comets and nebulæ.

CHAPTER IV.

THE ATMOSPHERE AND CLIMATE.

We may now pass from the standpoint of the astronomer to that of the physical geographer. Hitherto we have regarded the earth as a relatively very small body indeed among the innumerable hosts of space, but now we have to consider it by itself, as the confine of all possible human action and so far as we certainly know the sole theatre of the vast series of phenomena which we speak of as life. For our purposes it will be most convenient to follow a rather artificial division of the subject, to deal first with that outer terrestrial envelope, the atmosphere, then with the ocean, and then with the internal condition of the earth; a number of points concerning the crust will follow, and we shall add a short account of the distribution of life in space and time. Very much indeed of what we have to consider now is purely controversial, and as much as possible we shall write this in its natural, most interesting, and most valuable form, as discussion. It will be our endeavour to present the various views of the questions dealt with, not only with entire impartiality but without any original comment of our own, since an attempt to contribute to a moot question here would not only be out of place but might also prove to be greatly to the student's disadvantage. And having premised so much we may proceed to the consideration of several questions affecting the atmosphere and climate.

We may assume that the reader is already familiar with the chief facts of the **composition of the atmosphere.*** It will, however, be well to consider here the variations which this composition incidentally undergoes. The only very variable constituent is water and that we will deal with under rainfall. The relative proportion of *oxygen* and *nitrogen* varies slightly as has been shown by Bunsen and Regnault. The mean composition by volume is:—

Nitrogen 79·04
Oxygen 20·96

* *Elementary Physiography.* Chapter XVII.

but the last named investigator has noted in the port of Algiers a proportion of oxygen as low as 20·395.

Air from the gulf of Bengal has had a percentage of oxygen as low as 20·46, and from the mouth of the Ganges near Calcutta, 20·387. It is suggested by Thorpe,* however, that organic matter drawn into the tube of the aspirator might be partially oxydised and so falsify the analysis. Country air has shown a maximum of 21 per cent. The amount of oxygen in the air may be undergoing a very gradual secular reduction since the process of oxydation at the earth's surface must be considerably in excess of the reduction due to the activity of the chlorophyll in plants.

Ozone is a variable constituent of air, being least of course in the presence of organic matter, that is, in marshy districts and in towns, where it is scarcely to be detected at all. In country air, according to Thorpe, quoting Houzeau, the proportion is about 1 in 700,000 vols. The connection of an increased amount of ozone with disturbances in the electrical equilibrium of the air, though probable, has not been quantitatively determined. H_2O_2 is also present in country air and has been detected in rain, hail, and snow.

Carbon dioxide increases in amount during foggy weather even up to ·009 per cent. (volumetric). There is also a greater proportion at night since during the darkness the process of its assimilation by plants is at a standstill, and for the same reason country air has less than that of towns. Over the sea the proportion of carbon dioxide may fall to ·003 per cent. The total proportion produced by fires and the respiration of living things is according to Poggendorf, only about one-tenth of what is given off from subterranean sources. It must be borne in mind that a constant relation must exist between the atmospheric CO_2 and the amount in solution in the ocean. The gas is soluble to the extent of 1·8 vols. in one volume of water at 0° C., at the normal pressure, and proportionately more so as the pressure is increased. Any cause increasing the amount of this gas in the air must also increase the proportion dissolved in the oceanic waters, and conversely the withdrawal of carbonic anhydride from the atmosphere will involve the release of a corresponding quantity from solution in the sea.

Nitric acid in the air is probably formed (1) by electrical discharges through air, and (2) by the action of ozone on

---

* Thorpe's *Inorganic Chemistry.*

ammonia. The amount (probably due to the latter cause in this case) is greater over towns. In Glasgow it is as high as ·0002436 per cent.; in country places as low as ·00005. The first portion of the rain of a thunder shower always has an acid reaction due to this constituent. *Ammonia* derived from the decay of organic matters is a variable constituent, always present to some extent in the atmosphere.

Besides these constituents we find in the air the spores of bacteria (= microbes, germs), fungi, and algæ. We also find organic compounds due to animal and vegetable decay, sulphuretted hydrogen and other sulphuretted compounds. Over towns where coal, which invariably contains iron pyrites, is burnt, SO_2 occurs, and as a consequence of its further oxydation free sulphuric acid. Free acids do not occur away from localities of considerable combustion. Sulphates are, however, present in almost all rain water.

Clouds, Mists, and Rainfall.—Coming now to the water contained in the atmosphere. Water exerts a perceptible vapour tension even when solid. Its vapour tension at 0° C. is 4·6 m.m., and at − 32°, 0·3 m.m. So that snow and hoar frost disappear into the atmosphere even when the temperature is far below the freezing point. At 10° the tension is 9·2 m.m., and at the boiling point (100°C.), of course, 760 m.m. The measurement of the amount of water in the air is **hygrometry.** The wet and dry bulb instrument is described in the elementary work in this series.

Daniell's hygrometer is the simplest means for the direct determination of the dew-point (Fig. 12). Two bulbs, A and B, communicate. B contains liquid ether, and ether vapour fills A and the communicating tube. The bulb A is covered with linen, and B is made of black glass. Inside B is a small thermometer partly immersed in the ether. Ether is dropped upon

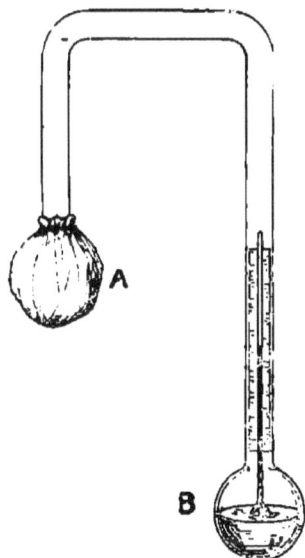

Fig. 12.

A; the ether vapour inside is thereby condensed, and evaporation goes on from the liquid in B. Evaporation causes a lowering of temperature, and by continuing the operation, B is eventually cooled down, until a film of moisture is deposited upon its surface. The temperature then indicated by the enclosed thermometer is the *dew-point*. The dew-point can also be indirectly obtained from observations with the *wet and dry bulb hygrometer*.

The instrument of Dines (Figures 13 and 14) consists of a cylinder (A) filled with a mixture of ice and water, communicating through a cock (B) with a pipe lying under a black glass slab (E). The pipe has a thermometer (C) inserted above it. The method of determining the dew point is as follows. The cold water from the cylinder is admitted to the pipe; the temperature of this falls, and so soon as the dew-point is reached the black glass is rendered misty by water condensed from the air. The thermometer is then read. Now by closing the cock, the water

Fig. 13.   Dines' Hygrometer.

in the pipe is cut off from its cold source and the temperature of the black glass rises until at the dew-point, the cloudiness evaporates again, and the thermometer is again read. This is a thoroughly practicable instrument. Human hair and catgut expand and contract with variations in hygrometric condition, and have been utilized in more or less scientific hygrometers. Seaweed becomes damp in moist air, by the action of hygroscopic salts in its substance.

Fig. 14.   Section of Dines' Hygrometer.

The distribution of water vapour in the air is extremely irregular, and so any estimate of the proportion of this constituent must be necessarily open to the possibility of considerable error. It is of course greatest near water surfaces, and this being allowed for, the lines of equal vapour pressure have a general coincidence with the isothermals. The vertical distribution of moisture however does not follow the upward diminution of temperature, the fall of the water constituent being more rapid. The proportion actually decreases more rapidly than would be the case in an independent atmosphere of water existing under its own pressure.

We may assume the reader familiar with the general facts of the **precipitation of water** from the air. There appears to be a cyclic recurrence of seasons of maximum and minimum rainfall, and this recurrence is, according to certain investigators, related to the cyclic occurrence of sunspots. A similar relationship has been supposed to exist between disturbances of atmospheric electricity and sunspot cycles. Maxima of summer rainfall appear to recur one or two seasons after sunspot minima, and maxima of thunderstorms about the period of sunspot maxima. This has been enforced by extremely convincing diagrams, but at the best the facts established are only imperfect coincidences. No direct causative connection at any rate has been shown between one set of pheno‑ mena and the other, though *a priori* we may assume that if the quantity of sunspots observed has a relation to the amount of radiant energy given out by the sun, a quantitative difference in terrestrial meteorological phenomena of some sort *must* follow, since the primary cause of all such phenomena is solar radiation.

Among the heaviest rainfalls recorded in one day is that of February 3rd, 1893, in Queensland, amounting to 35·7 inches. The highest record extant is that of Chirapunji (Cheerapongee) in the Khasia hills, where on June 14th, 1876, there fell 40·8 inches in the twenty-four hours. From the 12th to the 14th inclusive there fell altogether 102 inches. The Queensland fall, however, was quite abnormal for the district, and resulted in unexpected and therefore disastrous floods, while the Indian downpour, enormous as it was, was not so very much above the usual condition of things for that season of the year as to lead to any catastrophe. In England, one of the heaviest falls was one of 5 inches in Monmouthshire, July 14th, 1875.

There is little to add to what is contained in the elementary work regarding the formation and classes of *clouds*. With regard, however, to the exceptional blackness of thunder clouds, it has been shown by Shelford Bidwell that a jet of steam escaping gives rise, if electrified, to a very much darker cloud than when not electrified. This was explained by him as being due to the aggregation of the water drops into larger groups than before. Subsequent criticism, however, tends to the conclusion that the blackness is more explicable by assuming that the water drops are more finely divided through electrification than they were before.

Wells' time-honoured theory of the *formation of dew* has recently been called in question. Researches by Mr. John Aitken and the Hon. R. Russell have shown that this theory is insufficient to account for all the facts observed. The amount of dew frequently deposited is far in excess of the moisture that could be held in the surrounding air. This, therefore, led to the idea that dew is mostly formed by moisture derived from the ground and from vegetation. Among the experiments indicating that this is the case given by the Hon. R. Russell in his 'Observations on Dew and Frost,' are such as the following:—

A large quantity of dew was invariably found on clear nights in the interior of closed vessels over grass and sand.

Very little or no dew was found in the interior of vessels inverted over plates on the ground.

More dew was found on the lower side of a square, slightly raised, china plate over grass or sand than on the lower side of a similar plate placed upon the first.

The lower sides of stones, slates, and paper on grass or sand, were much more dewed than the upper sides. The flat wooden back of the minimum thermometer on clear evenings when lying on earth, sand, or grass, was almost invariably wet before the upper surface.

The lower sides of plates of glass, one or two inches above grass, were as much or more bedewed than the upper sides.

Leaves of bushes, leaves lying on the ground, and blades of grass were about equally bedewed on both sides.

The interior of closed vessels inverted on the grass and covered with two other inverted vessels of badly-conducting substance were thickly bedewed, and the grass in the three circular enclosures was also thickly bedewed.

The general conclusions of these observers do not tend towards the entire overthrow of the theory of Wells, but merely to the qualification that the moisture is not purely atmospheric in its origin as Wells contended.

We may supplement what is said in the elementary work
respecting fogs and mists by the distinction between wet mists
due to supersaturation and condensation on insufficient dust
particles, and dry fogs due to the bedewing of an excess of small
dust particles, as a consequence of their great radiation consequent upon the largeness of their surface relative to their mass.
In the latter case the fog particles coming in contact with warmer
objects do not deposit water but take up heat and are dried, while
in the former the amount of water relative to dust particles is so
great as, in view of the saturation of the air, to result in the wetting
of the body. (Compare *Practical Teacher*, November, 1893.)

Pertaining also to the question of the amount and condition of
atmospheric moisture is that of the *duration of sunshine*. This
duration is measured commonly by means of a large burning
glass, in the form of a sphere. This focuses the sun's rays upon
a circular piece of mill-board, which is burnt so long as the direct
sunlight falls upon the globe. The mill board is marked with a
scale upon the principle of the sundial, so that the duration of
hours of direct sunshine are measured by the charred trace.
Obviously this affords no measurement of the intensity of the
radiation.

Solar radiation is measured in an approximate manner by the
black bulb thermometer *in vacuo*. This is a maximum thermometer with a lamp-blacked bulb, enclosed in an exhausted glass
case. It is exposed to the sun's rays usually at about four feet
from the ground, and it is carefully protected from bodies which
might radiate heat to it. The outer case will evidently be at or very
near the temperature of the air, the temperature which would be
recorded by a thermometer in the shade. What is regarded as the
maximum of solar radiation is obtained by subtracting the
maximum of a shade thermometer from the maximum obtained in
this instrument.

Temperature of the Higher Air.—Our knowledge of the
condition of the higher air has recently been greatly extended by
the simple expedient of sending up a balloon of gold-beater's skin
filled with coal gas, and carrying a Richard registering apparatus
for temperature and pressure. An account of these (W. de
Fonvielle) appeared in *Nature* for June 15th, 1893. The balloon
was launched at Vauguard on March 21st, 1893, and was
recovered with the registering apparatus in good condition. The
registration of pressure came down to 95 mm., indicating a

G

probable height of 15,300 yds. The minimum temperature recorded was —51° C. The rate of diminution upward of temperature appears to have been a degree for 192 yards, a rate unexpected and rapid. Pure hydrogen balloons will probably by the time this is in print have followed from Meudon, and their records have been taken. The height attained will be checked by observations of the apparent diameter of the balloon, its zenith, distance, and altitude.

The rapid fall in temperature observed is in harmony with the supposition that the temperature of space is at the absolute zero (— 273° C.) or below it. At this temperature the gaseous constituents of the air would condense and drip back into the atmosphere. The condensation of air effected by Dewar has contributed to weaken the old belief in the univeisal applicability of Boyle's law, and the view that the void of space is occupied, even by gas of extreme tenuity. The belief is, however, stated in Thorpe's *Chemistry* (689) that the atmosphere extends through space.

Liquefaction of Gases.—*Liquid oxygen* has been obtained in the following way by Dewar. The chamber containing the oxygen to be liquefied was surrounded by two casings. Into the spaces between these casings and between the oxygen chamber and the inner of the two, liquefied ethelyne and liquefied nitrous oxide were introduced and evaporated, to pass to a compressor as a gas, and to be there recondensed and returned again to the casing space. The cycle of operations for each of these substances was generally similar to that employed in the case of refrigerating machines working with ether or ammonia.

The ethylene is obtained by the action of alcohol upon strong sulphuric acid ; the H_2SO_4, which has a strong affinity for water, dehydrating the alcohol.

$$C_2H_6O = C_2H_4 + H_2O.$$
Alcohol. Ethylene.

Ethylene can be liquefied by pressure at any temperature below 9·2° C. Its boiling point is —136° C., and by evaporating the liquid after cooling, extremely low temperatures are obtainable.

In this way oxygen has been cooled below its critical temperature ; that is to say, below the temperature above which it cannot be condensed by pressure. The critical temperature of oxygen is —112° C. At that temperature a pressure of fifty atmospheres is required to condense the gas. At — 136°, the boiling point of ethylene, the pressure needed is only 20 atmospheres. At − 182° it is only one atmosphere. In other words, the boiling point of oxygen is − 182°.

In his demonstration at the Royal Institution last May, Professor Dewar exhibited to the audience a pint of liquid oxygen. The liquid was at first cloudy, the cloudiness being due to the presence of impurities. After filtering, the element was seen to be a clear transparent liquid of a slightly bluish tinge. By keeping the vessel containing the liquid in a glass receiver over mercury at a low temperature, the evaporation is lessened, and it is possible to examine with some precision the physical properties of the liquefied oxygen.

The latent heat of vaporization of the liquid is about 80 thermal units. Its capillarity at its boiling point was about one-sixth that of water.

When the liquid was interposed in the path of the rays from the electric light, it gave an absorption spectrum substantially similar to that of the gas. The lines A and B of the solar spectrum, due to oxygen of the earth's atmosphere, came out very strongly. Altogether the liquid and the highly compressed gas show a series of five absorption bands, situated in the orange, yellow, green, and blue respectively.

(The fact of the oxygen lines of the solar spectrum being atmospheric, we may note here, is enforced by the sustained observations of Dr. Janssen on Mt. Blanc. The oxygen lines disappear more and more as a greater altitude is attained, and as the thickness of air through which the sunlight has to pass diminishes. And at sunset and sunrise, when the sun's rays pass obliquely through the air, they are seen more distinctly.)

The persistence of the absorption spectra of oxygen through all stages of its condensation, would seem to point to the fact there are no essential alterations in its molecular constitution during these changes.

Liquid oxygen is a non-conductor of electricity. It is strongly magnetic, twice as much so as a solution of ferrous chloride, nearly five times as much so as air. Its specific magnetism is $\frac{1}{5000}$ that of iron. Placed on a cup of rock salt — which it does not wet — it remains in the spheroidal state, and if the cup and its oxygen be then placed between the poles of an electro magnet, the element will be lifted out of the cup and will connect the two poles. It will then proceed to boil away off the poles, and when the circuit is broken it drops back into the cup.

Previous to Professor Dewar's recent work the chief names in connection with this question of the liquefaction of gases were those of Dr. Andrews, Cailletet, Pictet, Wroblewski, and Olzewski, so that Professor Dewar has recovered for this country in this line of investigation a supremacy it had lost.

Andrews investigated CO_2 chiefly, and was the discoverer of the *critical temperature*, that is, the temperature above which it is impossible to liquefy a gas. For carbon dioxide the temperature is 30·92°. Above this carbon dioxide tends to behave as a perfect gas, and at 48° it obeys Boyle's law (*i.e.*, volume inversely as pressure). A pressure of about 75 atmospheres liquefies this gas at the critical temperature. It is by observing the behaviour of the liquid inclusions in the minerals of rocks about this temperature that the liquid in many of them has been recognised as CO_2. At 31° the boundaries of the drop of liquid disappear and gas fills the cavity.

Cailletet compressed oxygen in a vessel surrounded by a freezing mixture. The pressure (of as much as 500 atmospheres) raised the temperature of the gas compressed. After this an interval was allowed to elapse during which this heat could radiate away from the apparatus. Then the screw of the press was suddenly released, and the pressure partially relieved. The gas expands, of course, and a great fall in temperature is the consequence of this expansion. In this way oxygen, nitric oxide, and marsh gas were liquefied, but only in small quantities.

Pictet obtained oxygen by heating potassium chromate which passed into a strong tube, the pressure upon the oxygen being produced by its own disengagement. He surrounded the condenser by carbon dioxide, which was evaporated by pumping ; this CO_2 being previously condensed and cooled by the evaporation of liquid SO_2.

Wroblewski and Olzewski were the first to employ liquid ethylene, and to obtain small quantities of liquid oxygen and air. It is an improvement of their method which has been employed by Professor Dewar.

Liquid air was obtained by Professor Dewar by the agency of the liquid oxygen above described. A temperature of below − 200° is attainable by the evaporation of oxygen accelerated by a high expansion pump, and atmospheric air may then be liquefied in an open test tube at the ordinary atmospheric pressure.

Liquid air is a murky blue liquid, and on its boiling in a test tube nitrogen is first disengaged and then oxygen. This is shown by the familiar experiment of inserting a smouldering splinter into the test tube from which the gas is being disengaged. At first the splinter does not re-ignite, but after an interval the whole of the nitrogen has gone and the oxygen in its turn beginning to assume the gaseous state, the splinter bursts into flame.

Liquid air, like liquid oxygen, is a non-conductor of electricity, and both constituents of the mixture are magnetic.

By evaporating liquid air in a similar manner to the above evaporation of oxygen still lower degrees of temperature are possible, and the attainment of the critical temperature and the liquefaction of hydrogen is thus very possible, and probably indeed imminent.

Nitrogen has a critical temperature of $-146°$, and its boiling point is $-194°$. We are evidently thus approaching more and more nearly to the absolute zero of temperature, $-273°$. At that temperature no gas, it would appear, can exist.

We may here remind the reader who is not especially a student of chemistry of Guy-Lussac's law of the expansion of gases ($=$ law of Charles). The reader has probably met with the statement that liquids and even solids evaporate at *all* temperatures—*e.g.*, gradual disappearance of snow during a prolonged frost—and he may be imbued with the old conception that an absolute vacuum is a physical impossibility. As we go down the scale of temperature the proportion of evaporation, however, steadily diminishes. Guy-Lussac has shown that for every degree upward or downward a gas expands or contracts $\frac{1}{273}$ of its volume at 0°. C. If we call its volume at 0° v, at 2° it will be $v + \frac{2v}{273}$, and at -2 degrees, $v - \frac{2v}{273}$. Evidently, then, at $-273°$ all the heat, all the latent energy of its state, will have left the gas, and its volume would be nothing; that is to say, it cannot exist at that temperature as a gas even under no pressure at all.

At these low temperatures there is a remarkable diminution of the activity of chemical reactions. We seem to be approaching what has been styled the *Death of Matter*. All who have studied chemistry must be aware of the strong affinity of potassium, sodium, and phosphorus for oxygen. Potassium, for instance, tarnishes at once on exposure to air, and, on being thrown upon water, decomposes that compound, to become oxydised with such violence as to burst into flame. But both potassium and sodium will float unaltered upon liquid oxygen. At $-200°$ the molecules of oxygen have only one half of their ordinary velocity, and have lost three-fourths of their energy. Professor Dewar has even stated that at these low temperatures all chemical action ceases. This, however, requires some modification, since a photographic plate immersed in liquid oxygen was still sensitive to light at −200° C. This is chemical action as a consequence of radiant energy. It would, however, appear to be correct to state that at these temperatures, in the absence of extrinsic energy, matter displays no chemical affinity.

Professor M'Kendrick, Professor Dewar stated in the conclusion of his lecture, had tried the effect of these low temperatures upon the spores of bacteria. Sealed glass tubes containing nutritive fluids were submitted to a temperature of −182° for one hour, and subsequently kept at a warm temperature for several days. On being opened they were found to contain abundant bacteria living. Seeds also resisted an equally low temperature. This, he thinks, has proved the possibility of Lord Kelvin's suggestion that life first reached this earth upon a spore-carrying meteorite. At any rate, it shows that the complicated chemical compounds which make up living protoplasm, though their activity is probably absolutely suspended, are not destroyed by these extremes. Simply nothing happens until the temperature is raised.

The general tendency of all these enquiries is to point downwards to a physical condition at which all phenomena cease, through the withdrawal of all energy in the form of heat from matter. And recent investigations into the temperature of the higher atmosphere seem to show that the temperature of interstellar space cannot be far above this absolute zero—this death chill of all material things. It is in harmony with the apparent absence of friction in stellar movements to suppose that interplanetary space is without gas even of extreme tenuity.

The Chief Variations of Atmospheric Pressure, periodical and in space, are to be accounted for when the necessary consequences of evaporation and changes in temperature are considered. Thus, during the winter the land in the north hemisphere is a region of high pressure, relative to the adjacent and warmer sea, and the reverse is the case in summer. The pressure around the south pole, however, appears to be abnormally lower than around the north, and the anti-trades blow with much greater violence in that hemisphere. Possibly this abnormality is due to the fact that there is apparently less land around the southern than the northern pole. The size of the Antarctic continent is not considerable. The

mean barometric pressure of the 60° of south latitude in July is
29·4 inches, in January it is 29·2 ; in the northern hemisphere the
lowest barometric pressure on the same parallel is 29·4 in the
North Atlantic, the greater proportion, however, of the air along
that line being at a pressure of 30·0   A somewhat contradictory
fact, however, to the view that this low, Austral barometric
pressure is due to greater circumpolar warmth is the greater range
of the southern icebergs towards the equator, and their more
considerable dimensions.

Atmospheric Movements.—The essential facts of the
atmospheric circulation are given in *Elementary Physiography*, but
we may add here some refinements of the explanation, some of
which have been re-stated by the late Prof. J. Thompson in
the Philosophical Transactions for 1892. The account given in
the elementary work is Maury's variation of Hadley's theory. As
a matter of fact there is not the slightest evidence that the 'anti-
trades' are really the return currents in the great atmospheric
circulation. Prof. Thompson is of opinion that the upper current
from the equator poleward remains the upper current until the
stagnant air of the North Polar Calms is reached. If the over-
current and undercurrent really did change places at the tropics,
instead of the Calms of Cancer and Capricorn we should
evidently have regions of extreme disturbance. The equator-
ward current, it follows, therefore, is beneath the poleward current
always. The south-west and north-west anti-trades Prof. Thompson
accounts for in the following way, and with the help of an
experiment.

If a shallow circular vessel be filled to a moderate depth with
water, and a small amount of suspended matter be in the water
to indicate its internal movements, and the contents of the vessel
be then set circulating by stirring, the small suspended particles
will be seen to pass along the bottom and collect at the centre of
the vessel.

They are carried there by a current determined towards the
centre by the centrifugal tendency of the lower strata of water
being diminished by bottom friction relative to the masses above.
The tray need not be motionless of course, so long as the
circulating motion of the water mass above is greater.

It was pointed out by Hadley long ago 'that the N.E. and S.E.
winds within the tropics must be compensated by as much N.W.
and S.W. in other parts, and generally all winds from any one

quarter must be compensated by a contrary wind somewhere or
other ; otherwise some change must be produced in the motion
of the earth round its axis.'

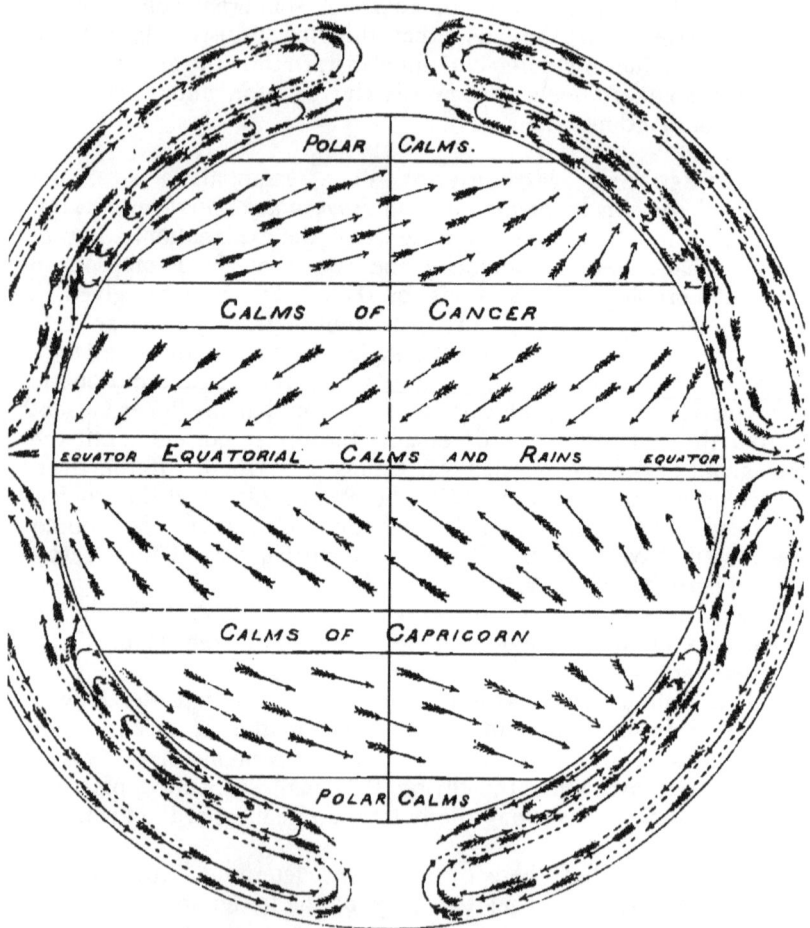

Fig. 15.   The Movements of the Atmosphere.

According to Prof. Thompson's reasonings, the Doldrum air should
rotate approximately with the earth.   As a result of observations
made in the spreading of the dust of the great Krakatoa eruption,

however, it has been inferred that in the upper strata of the air (13 miles high) over the equator, there is a *constantly blowing east wind*, having a velocity of about 70 miles per hour. Cirrus clouds have also been observed in the tropics in the Indian Ocean suggestive of a like current. This wind, if the observations cannot be questioned, throws a considerable amount of doubt upon all the various modifications of the great theory of Hadley that have been propounded.

Causes of Cyclones.—It will be well perhaps to give here a brief account of the views of the late H. T. Blanford and of Prof. Hann, of the origin of cyclones and anti-cyclones. Previously it has been assumed that these atmospheric conditions had a consequent relation to temperature variations in the lower strata of the atmosphere. Cyclones were connected with the condensation of moisture, the release of its latent heat, and the consequent rise of the warmed air. There can be no doubt that the temperature in a cyclone region is generally high, and that anti-cyclone weather is characterised by a clear sky and low thermometer. However, it would appear that these conditions of temperature and atmospheric moisture may be regarded not as causes but as effects of cyclonic or anti-cyclonic movements of the air.

It was shown by Prof. Hann that the low temperature of an anti-cyclone region was confined only to the lower strata of the air. In his memoir presented to the Vienna Academy in 1890 he gave the results of his study of an anti-cyclone lying over Central Europe from November 16th to the 24th, 1889. This extended over the observatories in the Eastern Alps. In the month before there had been a barometric minimum over the same region. Notwithstanding that it came six weeks later in the year, the mean temperature of the anti-cyclonic air column was more than 2° C. *higher* than that of the preceding minimum. The greatest depressions of temperature in the mountains in summer always occur during the cyclones, and Prof. Hann concludes that 'the warmth which accompanies winter cyclones at the earth's surface and which was assumed to characterise the whole column, is restricted chiefly to the lower atmospheric layers ; the observations on high mountains show that the greatest warmth is always brought by anti-cyclones ; the higher the mountain, the more pronounced is this result.'

Thus the 'upcast' theory of cyclones, so far as those of the temperate zone at least are concerned, falls through.

Prof. Hann explains the difference in temperature of the cyclone and anti-cyclone at the earth's surface as being a consequence of the warmth of the higher part of the anti-cyclone, its consequent capacity for water, and the clear skies which must necessarily follow. These clear skies offer little obstacle to terrestrial radiation, and the ground and the immediately superincumbent air are therefore cooled.

'Cyclones and anti-cyclones are but partial phases in the general circulation of the atmosphere.' The upper air currents flowing towards the poles are *converging* continually, continually being more 'congested' as they flow into circles of latitude with continually diminishing radius. This jostling together of the upper air currents is at first slight and increases as they approach the poles. Great masses of air descend, and the air streaming outward from these regions of downfall gives rise to cyclone eddies in the intervening spaces of less pressure. The downward conpression of the anti-cyclone column raises its temperature, except immediately at the surface where terrestrial radiation comes into play. One would expect from this that atmospheric disturbances would have their maximum when the higher or anti-trade currents are most active. They are most active in winter, when the temperature difference of polar and equatorial regions is at a maximum, and as a matter of fact the storms in higher latitudes are then most severe. Moreover, in years when the medium annual height of the barometer was below the average in the Indo-Malayan region, there was abnormial barometric pressure during the winter in Western Siberia.

The late H. T. Blanford, in his discussion of Professor Hann's work (*Nature*, November 6th, 1890), wrote, 'As regards the genesis of anti-cyclones, for the study of which the Sonnblick, Hoch Obir, and Sautes observatories have afforded him numerous opportunities, which he has turned to the best account, Professor Hann's conclusions appear to be unassailable. And in respect of the cyclones or barometric minima of the temperate and sub-Arctic zones, although the evidence is much less decisive and its conclusiveness may be challenged in some particulars, it must at least be conceded that his arguments are entitled to much weight; and the facts adduced greatly weaken, if indeed they do not altogether destroy, the validity of the views hitherto prevalent.' Mr. Blanford, however, objected entirely to the extension of Prof. Hann's explanation to the cyclones of the tropics.

In a subsequent paper (*Nature*, November 27, 1890), Mr.
Blanford gave a number of sound reasons for this objection, which
we may summarise as follows:—

(1.) The prime causes given by Hann, the lateral compression of
the anti-trades, is least in the tropics, since there the
circumference of successive circles of latitude diminishes
least rapidly. But cyclones may originate within 6°
of the Equator.

(2.) If cyclones originated in the higher atmosphere they should
from the first have a proper motion with the general
drift of the upper currents. Observation concerned
especially with cyclonic storms originating in the Bay
of Bengal shows that there is a period of what Mr.
Blanford styled 'prolonged incubation.' There are at
first slight squalls, which increase in number, violence,
and frequency. . The barometric pressure at the initial
point falls, and the unsettled weather is gradually
replaced by the typical cyclonic circulation. It is only
at this stage that the whole disturbance begins to move
bodily, drifting in a direction somewhere between N.E.
and W., presumably because it is only at this stage
that the ascending air becomes involved in the higher
atmospheric current. At first the motion is slow.
Icelandic cyclones, on the other hand, fit Hann's
explanation by moving from the beginning.

(3.) There is nothing in Hann's theory to account for the fact
that tropical cyclones almost invariably originate over
the sea.

(4.) If cyclones had their origin in the higher currents they
would not be broken up by hills. But on November 1,
1879, a violent and destructive storm was broken by the
low hills of Tipperah, and a considerable proportion of
the Coromandel cyclones are dissipated by the Ghats
and hill groups of the Carnatic.

The temperature of tropical cyclones at least is sufficient to
satisfy the 'upcast' theory.

Mr. Blanford is careful to distinguish the storms that traverse
Northern India in winter and early spring. These travel eastward
and probably originate in the manner suggested by Prof. Hann.

These cold storms of India occur as waves travelling from west
to east. They are preceded by a wave of high temperature and
accompanied by precipitation and cold; the cold being, it would

appear, proportional to the amount of precipitation that may occur. When the snowfall or rain is slight the cold wave may be transitory. Cloud and rainfall is especially characteristic of the eastern and northern side of the depression; to the west and south the skies are clear.

There can be no doubt that, if not generally applicable, the theory of a heated uprush must be occasionally used to account for some atmospheric disturbances even outside the tropics, since it is an adequate explanation and the conditions must at times obtain. It is doubtful whether any hard and fast line can be drawn between tropical cyclones, hurricanes, and typhoons on the one hand, and 'tornadoes' and whirlwinds on the other. Dust whirls and waterspouts are merely whirlwinds catching up loose material, an upward lifting being very characteristic—at least of the superficially smaller storms.

Mr. J. Lovel, for instance, has described several dust whirls seen on heated highways which he regards as simply miniature tornadoes. One he describes as occurring on May 11, 1893, whirled from right to left, moved slowly northward along a curvilinear track, and had the dipping up and down motion characteristic of many tornadoes. It dissipated into the upper air after a passage of about 300 yards. The rotation of these whirlwinds may be either right or left handed. A remarkable part about them, contrasting strongly with the typical cyclone of the temperate hemisphere, is their low vertical height. One, in the province of Nerike, in Sweden, on August 18th, 1875, while it wrecked houses and smashed trees, did not in the slightest degree affect clouds at even a low altitude.

The Electrical Conditions of the Atmosphere have been made the subject of a wide-spread enquiry by the United States Government in 1884—88, and the results of this form the substance of a voluminous report by T. C. Mendenhall, published at Washington. The staff of observers were organised and trained by Professors Cleveland Abbe, Rowland, and Trowbridge, and in 1884, stations were established at Washington, Baltimore, Boston, New York, Ithaca, and Ohio.

A special form of electrometer was employed, and water-dropping, mechanical and flame 'collectors.' The results on the whole are indefinite. One seems, however, sufficiently remarkable for especial notice. During an Aurora on May 20th, 1888, no unusual fluctuations in atmospheric electricity occurred.

The atmospheric potential is usually positive, and it has been thought that a negative change was indicative of bad weather. This coincidence does frequently occur ; but 'negative electricity in clear weather was observed at most if not all of the signal service stations on numerous occasions during the progress of the work. In many such cases precipitation occurred at points from ten to one hundred miles distant, but in others clear weather prevailed over almost the entire country. A number of instances of clear weather with negative potential occurred at Ithaca, where special attention was given to the matter by Mr. Schultze.'

The effect of dust, fog, smoke, and cloud in producing* negative potential has been noted by more than one observer. 'A fall of potential could certainly be predicted when a dust cloud was seen rising.' At Terre Haut, Indiana, on a day when a fog formed after sunset, the potential fell rapidly from + 1000 to −200 volts. Even a small cloud would send down a positive potential. Thus, at Boston on January 3rd, 1888, the potential had been steady positive, but about 11.30 a.m. a small cumulus cloud approached the zenith. The potential fell from + 32 to − 21, and rose again to + 6 as the cloud receded. Later, larger clouds appeared and the potential fell steadily.

High winds also caused a fall of potential.

The summer average potential is higher than the winter.

No definite laws were educed in the case of thunderstorms. Violent fluctuations are described ending in a rise of potential.

One other interesting result may be noted, and that is the extreme amount of local variations in potential. Instruments quite similar in every respect, separated by such an insignificant distance as 100 m. may give very different indications. Not only may the absolute measurements of potential vary in such a case, but the character of the fluctuations may also differ very widely.

Photographs of Lightning.—A considerable amount of light has been thrown upon the forms assumed by discharges of atmospheric electricity by means of photographs. Three kinds of lightning have usually been previously distinguished, viz., forked or crooked lightning ; sheet, heat, or summer lightning and a rare form known as ball or globular lightning. Photographs of light-

---

* Professor Oliver J. Lodge thinks this haze may be an effect of low potential instead of its cause.

ning flashes have shown, however, that many ideas as to the
character of lightning flashes are erroneous. In the first place the
zigzag dove-tail form in which lightning is generally represented
by artists does not exist in reality. Not a single instance of the
artists' lightning-flashes is found upon the photographs that have
been taken. The ordinary flash appears upon the pictures as a
stream of light following a sinuous or wavy course, and precisely
similar in character to a spark of an electrical machine. Other
kinds are classified as follows :—

(1.) Branched or ramified lightning, having much the same
      appearance as a map of a river into which a number of
      tributaries are flowing. To obtain an effect of this
      character, the terminals of the electrical machine must be
      separated so that the spark only just passes between
      them. Such flashes indicate that the discharge is in-
      complete.

(2.) Beaded lightning. In this kind a number of bright spots
      appear upon the flash. It can be produced experimentally
      by the discharge of a large quantity of electricity, hence
      it is probably produced when the quantity of electricity
      passing is much greater than in ordinary flashes.

(3.) Meandering or knotted lightning. In flashes of this kind,
      the electricity appears to have taken very circuitous
      paths, the stream of light sometimes having the form of
      an irregular loop or knot. The effect is probably due to
      an optical illusion, for the different parts of such loops or
      knots may be several miles apart, and only appear close
      together because they are in nearly the same line of sight.

(4.) Ribbon lightning. A large number of photographs show
      flashes of a flat or ribbon character. It is, however,
      believed that this appearance may be due to a movement
      of the camera. Moreover, two or more flashes often run
      along the same path in rapid succession, and their united
      photographs may form the band.

(5.) Dark lightning. Flashes appear sometimes black in pho-
      tographs instead of white. This is purely a chemical
      effect. If the camera lens is covered immediately after
      the photograph is taken this does not occur, it is the
      result of the action of a feeble diffused light after the
      flash has passed.

Past Variations in Atmospheric Composition.—It is not an uninteresting subject for speculation how far the composition of the atmosphere and atmospheric conditions may have varied in the past. Fossilized rain prints and hail markings, and the uniform progression of life in the past point to no catastrophic changes. We find what may be growth rings in the plant fossil *Nematophycus* of the Silurian, pointing perhaps to an alternation of seasons such as we experience now. There is however a remarkable absence of annual rings from the carboniferous vegetation. It has been imagined from the exuberance of vegetation that during the carboniferous epoch the temperature, the total humidity and the amount of carbon dioxide, may have greatly exceeded that obtaining at present. As regards the latter constituent, however, it may be pointed out that so far from its mere excess being favourable to plant life, it may soon pass beyond a most favourable proportion to a distinctly detrimental influence.

Neumayer, moreover, has pointed out that a very much larger amount of CO_2 than at present in the air during the carboniferous epoch would have resulted in a greater amount of the gas being absorbed in the sea, and therewith the capacity of the oceanic water for the solution of $CaCO_3$ would be increased, and the accumulation of such limestones as occur in the carboniferous formation would be rendered impossible.

There is a much greater probability that the temperature of the atmosphere as a whole may have fluctuated, and therewith of course the ratio of the water vapour in the air to the amount of water precipitated in ocean and sea. On the meteoric hypothesis this world, and perhaps also its atmosphere, must have cooled considerably since Palæozoic times, and it is believed by Mr. Jukes Browne, for instance, that the amount of water in the air was then relatively much greater and the mass of the ocean therefore much less than now. Similarly one would anticipate that during the cold of the glacial period the oceanic mass must have been more considerable, unless indeed the condensed water was not largely laid up on the land in the form of ice and snow fields. We find evidence of the submergence of the British Isles during this period to the extent of at least 15,000 feet, according to Prof. J. Geikie, or of 300 to 500 feet, according to other authorities. What is regarded as a raised beach occurs at Moel Tryfaen, in Carnarvonshire, 1,350 ft. above the sea, and one at 1,200 ft. near Macclesfield. But recently these higher raised beaches have been called in question by Mr. D. Bell, Prof.

Blake, and others. The Moel Tryfaen deposit is stated to have
the shells mixed as to their depths and habitat, and they are more
broken than typical beach shells. The two shells of bivalves
are also invariably separated. It is supposed by those who take
the lower limit of the submergence period that these higher so-
called beach deposits have been forced up into their present
position by glacial action. The evidence that the deposit at Clava
in Nairn (500 ft.) is really a beach, appeared as it was presented
to the British Association (Sept., 1893) to be remarkably strong.
(See p. 115.)

A period of great rainfall in later Tertiary times has been
hypotheticated by some geologists and named the Pluvial period. It
would seem to have been characterised by the formation of large
inland freshwater lake systems, of which we have the shrunken
and saline vestiges in the gradually evaporating salt lakes of
Western North America and Central Asia.

Climate in the Past.—Whether it is the case or not that
the atmosphere has varied very widely as a whole, it is certain
that in the past very extensive variations of local or climatic
condition have occured. We find, for instance, in the Arctic
region fossil remains of abundant floras, of Lower Tertiary,
Cretaceous, Jurassic, Carboniferous, and Devonian age. Taking
the successive geological formations in their order, we may show
the climatic changes they indicate in the following table :—

Past Climatic Conditions of the British Area.

QUARTERNARY PERIOD.

| Recent Formations | Temperate conditions growing warmer |
| Pleistocene   ... | Glaciers extending southward to the Thames, per-haps with milder 'interglacial' periods |

TERTIARY PERIOD.

Pliocene	Temperate climate
Miocene	Sub-tropical conditions
Oligocene	} Tropical
Eocene	

MESOZOIC.

Cretaceous	Sea of considerable depth.
Jurassic	Period of subsidence, corals and other indications of warmth
Liassic	Islands, humid air, abundant vegetation, warmth
Triassic	Arid land, large salt lakes, centres of inland drainage

PALÆOZOIC.

Permian	Similar conditions to the Triassic. (?) ice striated pebbles
Carboniferous ...	Warmth and abundant vegetation, corals in limestones
Devonian	Corals, warmth. Salt lakes and arid land to N.
Silurian	(?) Warm sea
Cambrian	Sea. (?) Warm.
Archæan	(?)

To come to precise figures it has been estimated by Mr. Clement Reid (*Natural Science*, 1892) that the winter temperature of the channel during the glacial epoch was at least 20° F. below that now obtaining, and that in the Mediterranean during the same period it was at least 5° below its present height. He thinks that the mean annual temperature along the line of the Thames could not have been much if at all above the freezing point. On the other hand the Eocene fauna and flora may point to an equal or greater range above present conditions. The evidence of intensely cold epochs similar to and prior to the glacial does not appear to be very strong. Striated rock fragments have been found in the Permian Breccia of Leicestershire, and these Prof. Bonney thinks may have been ice striated, but in no case is he absolutely convinced of this. The marks may be 'slickenside' striæ from faults. Certain boulders found in these Permian beds and the general appearance of the rocks incline him, however, to the idea that the Permian period was one of low temperature.

The probable causes of these variations in climatic condition have given rise to a very large amount of speculation. Among the more recent contributions are one by Professor Bonney to the *Contemporary Review*, Nov. 1891, and Sir R. S. Ball's *Cause of an Ice Age*.

The former discusses the matter from the point of view of physical geography. Practically there is no great difference between his conception of the necessary condition of British glaciation and that suggested by Mr. Clement Reid. He requires that the isotherm of 32° should coincide with what is now the isotherm of 50°. The lowering of the mean annual temperature 18° F., other conditions remaining constant, would bring back the glacial period, and would make the North of Scotland comparable to West Greenland at the latitude of Frederickshaab. The diversion of the Gulf Stream would however only lower the temperature of North Wales 7½° and of Cape Wrath 12°. Elevation might

H

directly reduce the temperature 3°—4°, and indirectly by rendering the North Sea a land mass, and making the climate less insular, a degree or so more in addition. But he considers such raised beaches as Moel Tryfaen satisfactory evidence that there was ice when the land was lower rather than higher, and he concludes therefore that geographical changes alone are not sufficient to account for the full rigours of the glacial epoch.

Sir R. S. Ball discusses the question from the point of view of the astronomer. He first disposes of the hypothesis that the sun is a variable star of long period, or that the variation of temperature of space due to the stars can be of very great amount. These are absolutely gratuitous hypotheses, and unless we altogether fail to account for the phenomena under consideration without them, there is no reason why they should be made. A similar objection attaches to the supposition that the earth's axis of rotation has changed. We can imagine no adequate cause for such a change. Prof. George Darwin calculated that the entire redistribution of land and ocean would only alter the axis of a rigid earth 3°. The main factor in climatic alteration is the precession of the equinoxes, coupled with the variations in the form of the orbit due to planetary attraction. With regard to the former, 'the total heat received by the earth from equinox to equinox while moving round one part of its orbit, is equal to that received while completing its journey round the remaining part.' If the line of equinoxes coincides with the major axis of the orbit the north and south hemispheres will have seasons of exactly the same length, but if the line of equinoxes is at right angles to the major axis, that hemisphere which has its summer solstice at perihelion will have a hotter summer of brief duration, and a winter longer and colder than the opposite hemisphere, and this difference will be at its maximum where these conditions coincide with a period of maximum orbital eccentricity. The seasons on the opposite hemisphere will at such a time be most equable. The cycle of the return of the line of the apsides to coincidence with the solstitial line is a period of 21,000 years (*vide Advanced Physiography*, p. 140). The variations of eccentricity occur in no clearly defined cycle. The hemisphere having an aphelion-winter under the conditions supposed will accumulate great masses of snow and ice towards its pole and will be further cooled by the increased radiation due to the precipitation of its atmospheric water. The intenser sunshine of the summer period will act on these accumulations under conditions similar to those under which brilliant sunshine falls upon the snow of high mountains at present. Much will be immediately reflected into space, much radiated immediately, and the rest will disappear as latent heat in the partial reduction of the ice cap. This, during the relatively brief period of warmth, will not be destroyed, but will pass on into the long winter portion of the orbit again to be there reinforced. Prof. Ball calculates that the maximum difference possible in the length of the seasons is 33 days. This will obtain at the maximum of orbital eccentricity which occurs at intervals of vast and unknown duration. While the orbital eccentricity remains extreme there will be an alternation of glacial and genial periods corresponding with the cycle of 21,000 years already alluded to.

Variations in the obliquity of the earth's axis would also reinforce this differential effect. Stockwell, however, has determined that the obliquity varies less than three degrees (from 21° 58′ 36″ to 24° 35′ 58″). Great value was given to this as a factor in climatic condition by Dr. Croll (*Climate and Time*), but it does not appear to appeal very strongly to Sir R. Ball.

The most striking contribution in Sir Robert Ball's book to this discussion is his correction of the view apparently implied in previous literature, that half of the heat received by a hemisphere is received during its summer and the other half during its winter. This indeed leads to the singular conclusion that in the nonglaciated hemisphere the winter was warmer than the summer. On mathematical grounds, which it would be out of place to discuss here, he considers that the amount of heat received by a hemisphere during its summer is to the amount of heat received in winter, as 627 is to 373 (when the obliquity is 23° 27′). He argues that a very hot summer of brief duration is an absolutely necessary condition to a glacial condition of things. Large glaciers imply sustained and heavy precipitation such as could not occur if the temperature was constantly low. A very hot summer would, however, be necessarily accompanied by winds in a northerly direction (compare the S.W. monsoons blowing towards Central Asia), and carrying an enormous supply of moisture.

We have already intimated that the evidence of glacial epochs previous to the great ice age is not very strong. Glacial remains have been described from Australia, of an age certainly before the Pliocene, and Dr. A. R. Wallace is inclined to regard them as Palæozoic, but the fact is not universally accepted. Even if glaciated pebbles really occur in some Palæozoic formations, they may have been derived from elevated regions, and are not necessarily proofs of extreme climate. Probably glaciated pebbles are accumulating now in the Himalayan region and in Switzerland. Moreover, all geologists are not agreed that there is evidence of genial interglacial periods such as the astronomical theory requires. They must have occurred, if that explanation is correct, at intervals of 21,000 years, since every 10,500 years the relative position of the hemispheres completely changed. Plant remains certainly occur among the Pleistocene deposits, but in New Zealand and North America abundant, and in the former case semi-tropical, vegetation extends to the very foot of a glacier. However, the astronomers seem disposed to insist upon the adequacy of their conditions, and to assert that ' even if there was no geological evidence ' of glacial periods, we should still be forced to believe they occurred from a consideration of the phenomena of precession, the revolution of the apsides, and the gradual variation of orbital eccentricity.

It was pointed out by Croll that as a consequence of the glaciation of one hemisphere, the centre of gravity of the earth would be shifted towards the pole around which the ice and snow are accumulated. The data for any estimate are entirely in the air, as may be seen by the results of different calculations of the resultant rise of the sea level in high latitudes. Dr. Croll makes this 80 feet, Mr. Heath 128, and the Rev. O. Fisher 409. Another question turns on the fact of the laying up of vast quantities of ice and snow upon the land. Some geologists imagine that this would diminish the volume of the ocean, but we have already thrown out the suggestion that these accumulations would rather be at the expense of the atmospheric water. It must be remembered also that the glacial period of one hemisphere would occur concurrently with an epoch of less condensation in the other.

Dr. Croll in his work also considers the interdependence of maximum eccentricity, maximum obliquity and the coincidence of the line of the equinoxes with the major axis, on the one hand, with the direction of the present Gulf Stream on the other. As under the glacial conditions the temperature differences between arctic circle and tropics will be much greater, the northern trade winds will be more powerful than at present. Similarly in the ' interglacial' southern hemisphere the difference between pole and tropic will be less, and the south-east trades feebler. *Assuming that the equatorial current is chiefly caused by the trades*, it will be shifted southward. If, now, a map of South

America is glanced at, the reader will see that such a southward shifting will first cause the stream to impinge upon and split at Cape St. Roque, and if carried further will result at last in the entire southward deflection of all its waters, and therewith of all its heat and climatic amelioration.

At present, as a matter of fact, the southern hemisphere has its shorter summer at perihelion, and its temperature differences are more extreme than those in our own quarter of the globe. The south-east trades are as a consequence more powerful than the north-east, and the equatorial current lies now to the north of the equator. This matter of the deflection of the Gulf Stream as a subsidiary factor in glaciation is not dealt with very fully in the newer work of Ball.

It will be noticed that in Sir R. Ball's view, it is admitted that the statement of definite periods of time for the recurrence of glacial epochs is impossible. Dr. Croll attempted to do this, and Prof. G. H. Darwin, while praising Sir R. Ball for his intimation of this difficulty, maintains that an eccentricity cycle may be stated at least in round numbers, and gives its length as about 200,000 years. The glacial epoch, he considers, may have occurred perhaps 100,000 years ago, on this assumption. Taking the lowest estimates of the earth's age, there should be in the record of the rocks the deposits of many more than a hundred successive glacial epochs, on this assumption. As a matter of fact there is no good evidence of more than one.

QUESTIONS ON CHAPTER IV.

1. Describe the methods which have been devised for recording the duration and intensity of sunshine on different days.

2. Give an account of the results which have been obtained by recent researches in connection with the origin of the dark clouds which prevail during thunderstorms.

3. Give an account of the results which have been obtained by recent research in connection with the origin of fogs and mists.

4. Describe the mode of formation of hoar frost.

5. Describe the character of the cyclonic and anti-cyclonic movements of the atmosphere north and south of the equator respectively.

6. What do you know of the supposed causes of the glacial epoch which have been advocated by different authors?

7. What is the dew-point? Describe apparatus for its determination.

8. Give a full account of local variations in atmospheric composition.

9. Discuss the two chief theories of the origin of cyclonic movements.

10. Give an account of the results obtained by recent research in connection with the physical conditions of the higher air.

11. Describe the general atmospheric circulation.

CHAPTER V.

TERRESTRIAL WATERS.

The Composition of the Oceanic Waters.—The reader will be already familiar with tables showing the percentage composition of sea water. With the exception of the lime the relative quantities of salts in solution is extremely constant. The salinity varies with the conditions of evaporation, the greatest oceanic salinity being from the North Atlantic about 23° North (3·7 per cent.), and the least from a point south of 66° South in the Indian Ocean (3·3 per cent.). The average is 3·4. The proportions of dissolved salts as given by Prof. Dittmar in the Challenger Report are as follows:—

NaCl	77·758
$MgCl_2$	10·878
$MgSO_4$	4·737
$CaSO_4$	3·600
K_2SO_4	2·465
$MgBr_2$	0·217
$CaCO_3$	0·345
	100·

The lime varies from 1·585 to 1·825 per cent., but no relation to depth or superficial distribution has been determined. Of course the elements specified do not exist in the actual combinations indicated above. All solution appears to be accompanied by dissociation. 'In sea water as in any mixed solution each base is combined with each acid, and as there are four bases and four acids there must be sixteen salts, the individual percentages of which we have no means of determining.' (Dittmar.) The magnesium, for instance, may be chiefly in the form of a double chloride with the sodium, and some also, and perhaps even much of it, is probably combined with the carbonic acid. Sea water is alkaline, the alkalinity being due to the carbonates present.

Of the gases dissolved in the sea water the oxygen and nitrogen are derived from the atmosphere, and their proportions depend in the case of standing water upon the conditions of atmospheric temperature and pressure at the locality of the sample analysed, and in the case of water moving in a current on a complex of the previous and present conditions of the current. The proportion of these gases varies of course with the depth and diminishes on the whole downward. The carbon dioxide present is greatest in amount towards the bottom. From the difficulty experienced in getting it out of the sea water it would appear that it is not free but in a state of chemical or quasi-chemical combination (bicarbonate? Dittmar). We have already alluded to the state of equilibrium which must always tend to obtain between the atmospheric and oceanic carbon dioxide (p. 76). Since CO_2 is less soluble as the temperature rises, surface water at the equator must give off this gas, while in the polar regions the reverse is the case. The sources of terrestrial carbonic acid are referred to in our chapter on the atmosphere. *A priori* Dittmar assumes that in submarine volcanic explosions occurring in the deeper sea, the carbonic acid given off will be liquefied, since the temperature is there below the critical point of this gas, and the pressure sufficient. No such 'springs' of carbonic anhydride, however, appear to have been hit upon by the Challenger soundings, though they may have been approached in such a case as water sample No. 1096, where 15·8 c.c. of CO_2 were found to the litre.

Although it seems natural to assume that *all* elements present on the earth's crust must be represented in solution in the ocean, yet as a matter of fact, according to Forchhammer, only the following in addition to those specified above have been described.

Iodine (very minute traces indeed, but present in abundance in seaweed).

Fluorine (in boiler crusts of Atlantic steamers and directly determined).

Phosphorus (in phosphates).

Nitrogen (besides occurring free, in ammonia and organic matter).

Silicon (as soluble silicates).

Boron (determined directly, present in some seaweeds).

Silver, Copper, Lead, Zinc, Cobalt, and Nickel (inferred from their occurrence in certain seaweeds and corals).

Iron (directly and easily determined).

Aluminium (directly determined).

Barium and Strontium (also sufficient for direct determination).

Arsenic.

Lithium (by its very characteristic spectrum).

Cæsium, Rubidium, and Gold.

It has been pointed out that while potash felspars predominate over those containing soda in rocks, the sea contains a larger relative proportion of sodium salts. It may be possible that the proportion of these two elements in sea water is undergoing a slow change due to the continual addition of a slightly larger proportion of compounds of the former element.

It is interesting to compare these results with the analyses of the water of certain inland seas. In the Caspian we have a sea cut off from the oceanic water by extensive earth movements and into which the supply of river water balances or even slightly exceeds evaporation. The salinity is on the whole slightly less than that of the ocean, but there are extensive local variations. Thus near the mouth of the Ural we have almost the composition of river water. The salinity being only ·63 per cent., and the salts having the proportion :—

$NaCl$	58·
$MgCl_2$	09·5
$CaCl_2$	00·2
KCl	01·2
$MgBr_2$	00·3
$CaSO_4$	07·7
K_2SO_4	02·7
$MgSO_4$	20·4
	100·

At Baku, however, where the currents are slight and evaporation is very extensive, there is a salinity of 13 per cent., and the proportion of salts is

$NaCl$	.·.	64·2
$MgCl_2$	...	02·3
$CaSO_4$	...	08·5
K_2SO_4		00·4
$MgSO_4$		24·7
		100·

The Great Salt Lake of America has a salinity of 15 per cent. Unlike the Caspian it is not an isolated portion of the general sea, but the dwindling vestige of a great system of inland drainage. The composition of its salt is interesting in connection with the supposition noticed above of an ultimate increase of potash, due to aerial denudation, in the oceanic waters. The analysis, according to C. D. Allen, gives roughly

$NaCl$	...	79·0
$MgCl_2$	...	10·0
KCl	...	00·5
$CaSO_4$	...	00·5
K_2SO_4	...	03·6
$MgSO_4$	...	06·4
		100·

The salinity of Dead Sea water is 26 per cent., and the analysis of its solid constituent gives the following proportions:—

$NaCl$	...	...	13·9
$MgCl_2$	...	...	61·3
$CaCl_2$	...	...	18·1
KCl	...		03·4
$MgBr$	...		03·0
$CaSO_4$	...		00·3
			100·

In this case the concentration of the water has reached the pitch of deposition. Most of the magnesium sulphate (gypsum) has been deposited and the sodium chloride is following. The calcium chloride is also all thrown down as the tributary streams enter this sea. There must therefore be forming beds of salt upon gypsum, with intercalary layers of clay representing the rainy season of spring, very similar to the triassic deposits of our own latitudes.

Oceanic Circulation.—The essential facts of the oceanic circulation will already be familiar to the reader. There are two chief views of its cause.

(1.) That the difference of temperature of equatorial and polar regions causes a difference of specific gravity, and a general movement of warmer and lighter surface water from the equator to the poles, and colder and heavier bottom water from the poles to the equator (Maury).

(2.) That the impulse of the prevailing winds on the surface of the oceans produces a movement of the upper layers which is transmitted through the entire mass of water (Herschell and Croll).

According to the first theory, the water at the equator is lighter than that in colder regions, and therefore stands at a higher level. There is thus a gentle slope from the equator to the poles, and this causes a general movement of the warm upper layers of water from low to high latitudes, and a counter movement of the cold under layers from the poles to the equator ; the direction of movement in each case being, of course, modified by the earth's rotation. Much can be said in support of this theory, for as long as the equator receives more heat than the poles, so long must tropical waters expand and become specifically lighter than those in the temperate and frigid zones, and so long must convection currents be set up and warm surface currents and cold under currents exist. Against it, however, must weigh the fact that the chief poleward currents are obviously so directed by the trend of land masses, and that there are large superficial currents towards the equator. Moreover, since water is not very diathermanous, and the water below the surface, even in the tropics, is not perceptibly above the polar waters in temperature, the difference in level between equatorial and polar waters will not be comparable at all to what occurs in the diathermanous air.

When the effect of the wind in causing waves and obstructing tides is considered, it is not difficult to understand that a constant wind may drive the water of the ocean before it. But though it is easy to understand the explanation of surface currents on the wind-impulse theory, it is not so easy to understand how the motion of deeper water is affected, and the deep sea circulation kept up by such a movement. This view, however, is held by many eminent geographers, and had the strong support of the late Dr. Croll. It has been pointed out that much of the difficulty experienced in comprehending how under currents are produced by the wind arises from a deception of the imagination. Though an average depth of say three miles gives a striking impression, when the vast area of the oceans is also taken into consideration, shallowness rather than depth is the impression produced. In fact, the oceans of our globe are strictly comparable to a sheet of water 100 yards in diameter and an inch deep. As soon as this is recognised, there is no difficulty in understanding how the movements of surface water, brought about by impulses of prevailing winds, are communicated by the friction of successive layers down to the water at the ocean bottom. There is moreover, experimental proof for the view. Mr. Clayden has constructed a kind of Mercator's projection map of the world in relief with means for the imitation of the trade winds by jets of air. In this way a very good imitation of the surface currents at least of the ocean is obtained.

A subsidiary theory of oceanic circulation is that unequal evaporation at different parts of the globe causes a difference of saltness of ocean waters, and therefore a difference of specific gravity. There is, therefore, a sinking of the denser water to a lower level and a movement of lighter and fresher surface water to take its place. That such an action is sufficient to produce movements of water is shown by the Gibraltar current. But the amount of evaporation that occurs at any place is determined to a large extent by the character of the prevailing winds, hence, in all probability, the effects due to evaporation are primarily caused by winds. The effects of this cause must evidently be exactly the reverse of those given under (1).

However, the local difference of evaporation is a cause which *must* operate. Possibly the constant cold deep water current from poles to equator has no direct sequential connection with the system of surface currents, gulf stream, equatorial and the rest, but is caused by the constant transfer in the tropics of water from ocean to atmosphere and its return to the sea in higher latitudes.

The general facts of submarine deposition need not detain us now, but we may notice here some important investigations upon the conditions of accumulation in the deeper sea.

Deep Sea Deposits.—An account of the deep sea deposits based on the Report of Messrs. Murray and Renard will be found in the *Elementary Physiography*. The material of these deposits is derived (a) from the remains of pelagic organisms (*i.e.*, surface life,--foraminifera, radiolaria, diatoms, heteropods, pteropods and algæ chiefly, whales' ear bones, sharks' teeth, &c.) and of abyss organisms (echinoderms, annelids, &c.) which feed upon the sinking remains of the former; and (b) from mineral substances of terrestrial or extra terrestrial origin. Probably the greater part of the oozes has passed through the alimentary canals of the deep sea animals, but it is the pelagic forms which constitute the mass (90 per cent.) of these deposits. Albuminoids and other organic matters are present in them all, and the decomposition of these substances gives rise to H_2S and so leads to the formation of sulphides, particularly of sulphide of iron in the blue clays. Radiolarian ooze has not been described from the Atlantic and diatom ooze only from the antarctic or arctic circles. Pteropod ooze is found only in the Atlantic. The algæ that contribute are calcareous algæ, and their vestiges are small stud-like bodies and rods called coccoliths and shabdoliths. Otoliths (ear stones) of fishes occur, but it is a very remarkable thing that the skeletons of these and of the higher crustacea are conspicuously absent. The ear ossicles of whales also occur. Probably the large amount of organic matter and phosphate of lime in the crustacean and vertebrate skeletons contribute to their solubility. Sponge spicules are frequent but never very abundant. The six rayed forms (hexactinellidæ) occur in the deeper waters, the four rayed (tetractinellidæ) and straight forms (monaxonidæ) in the shallower areas. Of the fish teeth dredged some have been referred to extinct species of Pliocene age, lamna, otodus, and carcharodon.

The mineral substances of terrestrial origin found in the abyss deposits are possibly entirely volcanic, except where ice drift may have introduced terrigenous elements. Pumice lumps are frequent, amid finer fragments of glass and volcanic minerals. Perhaps on the whole a basic glass is most abundant, but acid and intermediate fragments occur. A considerable proportion of the pumice may be derived from subaërial volcanoes and have floated over the deep sea areas before becoming water-logged and sinking, but much of this volcanic material is doubtless ascribable to submarine eruptions.

There are also two types of constituent which are regarded as being of extra terrestrial origin. These are (a) the black and (b) the brown spherules which occur universally in pelagic deposits, but in the greatest abundance (25 to a quart of deposit) in the red clay of the Pacific where deposition has probably been slowest. (Figs. 16 and 17.) The black spherules never exceed one-fifth of a millimeter in diameter. They are almost perfectly spherical save for one slight flatness or depression which is universally present. They have the constitution of siderites (iron meteorites), a nucleus of iron or iron with cobalt and nickel, and a crust of magnetic oxide of iron. The brown particles are at least twice the size of these, magnetic, and consisting chiefly it would seem of some indeterminate monoclinic pyroxene, the magnetism being due to blackish brown inclusions therein.

There is no denudation process at work on these accumulations, but numerous slow chemical changes appear to be in operation. The red clay, which is estimated to cover nearly half of the

Fig. 16.   Black Magnetic Spherules.

abysmal area is, Messrs. Murray and Renard consider, derived from the decomposition of the volcanic materials already specified. The pumice lumps mentioned above are usually superficially changed to a clayey material, the basic glass is often altered to a brownish resinous substance like palagonite.   Minute crystals of phillipsite are abundant.   It is possible, however, that a proportion of the red clay may be terrigenous.

Fig. 17.   Brown Spherule.

Nodular masses of hydrated oxide of manganese (Fig. 18), mixed with earthy material and limonite, are also locally abundant.   They are concentric accumulations round such nuclei as sharks' teeth and pumice fragments. The hydrate is non-crystalline. Dr. Gibson has also found tellurium, lead, copper molybdenum, thallium, vanadium, zinc, nickel, and cobalt in these concretions. Mr. Murray considers the manganese derived from the volcanic material, but M. Renard imagines it may be thrown down in some way from the sea water. Mr. Teall, quoting Dieulafait, suggests that it has been precipitated from a solution of the carbonate through oxidation. It has also been suggested by

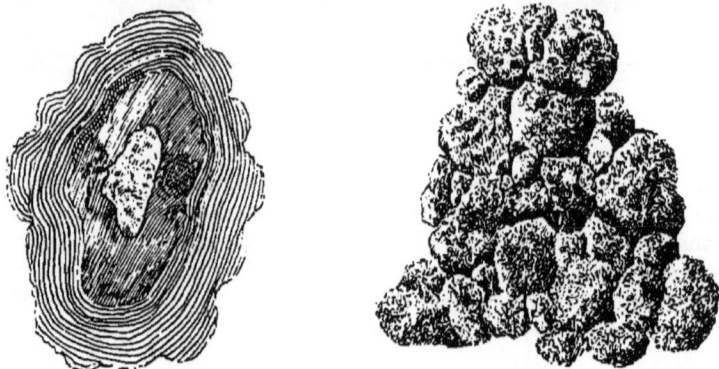

Fig. 18.   Manganese Nodules.

Buchanan that the manganese sulphate in sea water is reduced to sulphide by decaying organic matter and then oxidised.

Glauconite occurs chiefly in the green sands and muds of the less profound deposits, and the grains appear to begin to form within foraminifera shells, which they rupture, and subsequently to grow after the disappearance of the shell.   Precisely similar glauconite grains occur in the English greensand.

Bearing upon the question of the permanence of oceanic areas is the fact that calcareous constituents of oozes diminish rapidly below the 2,000 fathom line.   Compare the table in the *Elementary Physiography*, p. 267.   Coccoliths, rhabdoliths, heteropods, and pteropods do not occur in high latitudes, and the foraminifera are smaller.

The terrigenous deposits (*i.e.*, non-pelagic deposits below the 100 fathom line, which is roughly the boundary of the continental plateaux,) are described in the elementary work of this series. The green sands are glauconitic, the blue coloured by organic matter and sulphide of iron. The latter smell of H_2S, and both are widely distributed. Red clay occurs along the Brazilian coast, covering perhaps 100,000 sq. miles, and is coloured by an ochreous mineral. Volcanic muds and sands occur round volcanic islands, the latter where there are currents to remove the finer particles.

General Oceanic Conditions in the Past.—In the past it has been assumed by Professor Darwin that the rotation of the earth was more rapid than at present, the ellipticity of the earth's figure greater, and the tidal action of the moon more powerful. He has derived his data especially from the consideration of the retardation and heat that must be produced by the superficial friction of the tides, and to the internal friction due to the fact that the earth is not perfectly rigid. He concludes that 46,300,000 years ago there was a sidereal day of $15\frac{1}{2}$ hours, and that the moon was only at two-thirds its present distance, and 56,800,000 years ago the day was $6\frac{3}{4}$ hours, and the moon less than a sixth of its present distance. The lunar month, at that time, lasted a day and a half. Evidently the world lived a faster life in its earlier days. Oceanic denudation, due to tidal scour, must have been more rapid, and the trade winds and therewith the ocean currents must have been more powerful. Since temperature differences were more extreme and change more rapid, storms must have been more violent. However, as Professor Archibald Geikie points out in his Text Book, there is, on the geological side, no evidence in the older sediments (? Archaeans) of such differences in the condition of sedimentation.

We have already considered, in connection with the condition of the atmosphere, the question of variations of climatic condition. We may now direct attention to a wider question of a similar nature, and that is the past topography of the world. Beginning with the widest question, a very considerable amount of literature has accumulated in the discussion of the **'Permanence of the Great Oceanic Areas.'** By this phrase, of course, it is not intended that the ocean has had continuously the same boundaries but, as recently redefined by Doctor A. Russell Wallace, that the oceanic areas beneath the 2,000 fathom line, constituting about 70 per cent. of the whole ocean, have been ocean '*throughout all known geological time.*'

Among those who have advocated the view of this permanence are Dana, Darwin, Ramsay, Sir Archibald Geikie, the Rev. Osmund Fisher, and Mr. John Murray, and their unanimity of conclusion, while starting from very divergent bases, is advanced by Dr. Wallace as in itself an argument for his view. On the other side we have the late Sir Charles Lyell, Mr. Jukes Browne, Dr. Blandford, and Professor Suess. Some of the most important considerations in the matter we may summarise here.

i. The land surface of the globe is estimated by Mr. Murray as 28 per cent. of the whole superficial area of the world; its average height as 2,250 feet. The mean depth of the oceans, disregarding land-locked seas, is about 15,000 feet. The diagram we give (Fig. 19) is adapted from that of Dr. Wallace, and shows in its

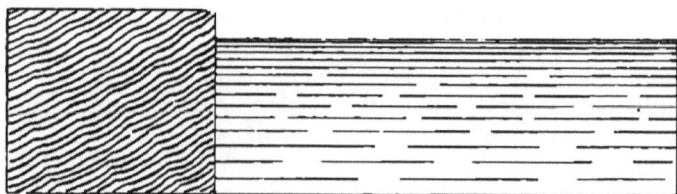

Fig. 19.   Proportion of Land to Water.

simplest form the land standing out of the sea.   He argues that for a new and large mass to rise out of the oceanic depth by elevation, to replace a continent, without compensating depression of other parts of the ocean bed, would simply involve universal submergence. Though such a coincidence might happen once, yet if earth movements of such magnitude did occur, the occasional total disappearance of all land would be a probable thing, and the disappearance of whole continents before others rose to replace them a highly probable event.   It is, however, suggested on the other hand, that in Palæozoic times the temperature may have been generally higher, the amount of terrestrial water in the atmosphere greater, and therefore the mass of oceanic water relatively smaller.

ii. Next it is pointed out that the contours of the ocean bed, as revealed by soundings, are not like the contours of the land surface.   We fail to find precipices and mountain ranges and peaks, and oceanic islands run sheerly down and present none of the ravines and radiating spurs and valleys which would occur on a submerged mountain. Most remarkable and significant is the absence of indications of geological faults.* On land a fault is usually not apparent by any change of level at the surface because subaërial denudation has kept pace with the formation of the fracture. But under the sea, where faults would be protected from atmospheric wear, they would appear, if they appeared at all, as great cliffs as high as their downthrow. Now deep sea soundings have failed to find such cliffs (cf. Gregory's *Elementary Physiography*, p. 260), and we are forced to infer, therefore, that movements of upheaval and dislocation, such as have formed the continents, have not occurred in the ocean bed. In the words of the Rev. O. Fisher, 'the compression which has caused the thickening accompanied by corrugation, such as characterises most elevated tracts, is properly a continental phenomenon and has no analogue beneath the ocean.'†

iii. All continents and continental islands present the same range of geological formations from Archæan to Quaternary, and such formations are indicative in almost all cases of the near presence of land.

iv. There are, it has been argued, no sedimentary rocks in oceanic islands, but they occur in the Seychelles, New Caledonia, South Georgia Archipelago, and New Zealand. Moreover, Mr. Crosby has pointed out that if existing continents were submerged the mountain peaks remaining above water would be mainly eruptive (? plutonic) rocks.

v. It has been asserted that there are no abyss formations‡ with the one possible exception of the chalk on continental land. The chalk, however, Dr. Wallace supposes to have been formed in a sea of only moderate depth. Since this argument was advanced earthy deposits have been found in the West Indies representing radiolarian ooze, globigerina ooze, and red clay. In Barbadoes and Trinidad similar deposits occur. Radiolarian earth is found in Algeria and foraminiferal earth in Malta. Dr. Hinde has described radiolarian charts from the south of Scotland. Moreover simply varietal and specific differences divide the oceanic foraminiferal ooze of to-day from the chalk foraminifera. The differences in composition that certainly do occur between chalk and recent foraminiferal ooze are to be explained by the supposition that the Cretaceous sea was warmer than the

---

* *Elementary Physiography*, Chap. XIII. †*Physics of the Earth's Crust.*
‡ *Elementary Physiography*, pp. 266—7.

I

present Atlantic, and by the fact that the lamellibranch *Inoceramus*, comminuted fragments of whose shells constitute a large proportion of the chalk, is now extinct. It is alleged that oceanic calcareous ooze has a higher proportion of silica than chalk, but this may be due to the segregation of flints in the latter. M. Cayeux supports the view of Dr. Wallace that the chalk is a shallow water deposit. He bases his arguments largely on the small derived minerals of the chalk. Dr. Hume, however, as the result of his recent investigations (July, 1893), maintains the deep-water origin of this deposit, and ascribes these minerals to the portative power of ice carried by currents, or they may in some cases have been brought to the deep sea in the intestines of fish.

vi. It was advanced by Dana that the great oceanic troughs marked regions of maximum radial contraction of the earth, and must therefore be, *a priori*, permanent. But this is opposed to the views of the Rev. O. Fisher on the internal condition of the earth, and those who argue against the view of permanence confront the one view with the other.

The nett tendency of the controversy in England is towards a compromise. Dr. Wallace remaining in possession of his 2,000 fathom boundary, at least as far as Mesozoic and subsequent ages are concerned, but abandoning his previous position of 1,000 fathoms. This drift is illustrated by these words of Mr. Jukes Browne—' I feel convinced that the truth is neither with those who assert the complete permanence of oceans or continents nor with those who teach the frequent conversion of one into the other. Similarly, that the existence of the calcareous ooze in modern oceans should be regarded as proving the present to be only a continuation of the cretaceous period I hold to be as untrue as the opinion that the chalk is not a product of a cretaceous *ocean*. So also with the arguments of those who urge, and those who oppose the doctrine of uniformity in the rate of geological change ; the uniformitarian may push his advocacy to such an extreme that he departs almost as far from the truth in one direction as the convulsionist does in another. Avoiding these extremes we may believe in the long continued existence of continents and oceans, and yet admit that the chalk is a genuine oceanic deposit ; we may adopt the doctrine of uniformity as our guiding principle in the interpretation of the past and yet believe in the theory of evolution.'

The views of Prof. Suess however are, as he states them, directly opposed to this compromise. Like Dana he regards the oceanic troughs and continental ridges as the result of the secular contraction of the earth, but unlike Dana he doubts their antiquity. In Pre-palæozoic times he thinks there was no land and only a shallow ocean, 'a universal hydrosphere or panthalassa '; and both the extreme heights and depths, he seems to hint, are new features of the globe due to its continuous cooling. He points out that the first fossil forms are those of shallow water organisms,* and that the anatomy of all land inhabitants and all deep sea creatures is regarded by zoologists as modified from that of shallow-water inhabiting ancestors. Moreover, he quotes the practical identity of European and Jamaica cretaceous fossils, and the fact that until the Quarternary period the molluscan types of the Chilian coast and of the Mediterranean were the same, in support of the view that a coast line 'may-be only an interrupted line' stretched across the present Atlantic and enabled these creatures to so distribute themselves. 'The Mesozoic deposits of Chili and those recently found at Taylorville in California show purely European characters, and the Neocomian of Bogota is the exact equivalent of that of Barreñre.' The views of Suess may be taken therefore as practically antithetical to those of Dana who asserts that the North American continent has grown around a nucleus of Archæan age. The English anti-permanence party do not appear to go so far as Suess but merely doubt the existence of the great oceanic troughs before Mesozoic times.

In a number of places, beaches and sea-worn caves are found at a considerable distance above the high-water mark of the neighbouring sea (see p. 96). For nearly a century the view has been held that such beaches and caves have been 'raised' above sea-level by internal disturbances. But the new school of geologists, under the leadership of Prof. Suess, of Vienna, have revived the old idea that the beaches and caves in question are left high and dry by the retreat of the sea, and do not represent results of upheaval ; in other words, the water is said to leave the land, and not *vice versâ*. Changes in coast-lines would thus be the result of **secular movements of the sea**. It must be conceded, however, by all geologists, that the cooling of the earth has caused distortions in the crust, hence there is little doubt that the general trend of coastlines must have been changed by movements of the lithosphere.

---

* But some of the blind and large-eyed trilobites of Bohemia resemble in these respects living deep water crustacea.

Prof. James Geikie has stated the cases of the old and new geology as follows:—'Geologists have maintained that the mysterious subterranean forces have affected the crust in different ways. Mountain ranges, they conceive, are ridged up by tangential thrusts and compression, while vast continental areas slowly rise and fall, with little or no disturbance of the strata. From this point of view it is the lithosphere which is unstable, all changes in the relative level of land and sea being due to crustal movements. Of late years, however, Trantschold and others have begun to doubt whether this theory is wholly true, and to maintain that the sea-level may have changed without reference to movements of the lithosphere. Thus Hilber has suggested that sinking of sea-level may be due, in part at least, to absorption, while Schmick believes that the apparent elevation and depression of continental areas are really the results of grand secular movements of the ocean. The sea, according to him, periodically attains a high level in each hemisphere alternately, the waters being at present heaped up in the southern hemisphere. Professor Suess, again, believing that in equatorial regions the sea is, upon the whole, gaining upon the land, while in other latitudes the reverse would appear to be the case, points out that this is in harmony with his view of a periodical flux and reflux of the ocean between the equator and the poles. He thinks we have no evidence of any vertical elevation affecting wide areas, and that the only movements of elevation that take place are those by which mountains are upheaved. The broad invasions and transgressions of the continental area by the sea, which we know have occurred again and again, are attributed by him to secular movements of the hydrosphere itself.'

It has been advanced by the Rev. O. Fisher, and has received a considerable amount of support, that the oceanic mass is on the whole increasing through the disengagement of vast volumes of steam in volcanic eruptions. On the other hand, Lockyer and others have supposed that the constant inward percolation of water and its combination in the hydration of minerals, *e.g.*, in the formation of talc, serpentine and hydrous mica, is sufficient for the supposition that the ocean tends towards ultimate complete absorption into the terrestrial body.

Variation of Latitude.—In recent years considerable attention has been drawn to the fact that the latitude of a place, as determined by observation of stars, is subject to a periodic variation. The difference in the observed positions of stars from which the latitudes were deduced could be produced by (1) a real periodic change in the direction of the earth's axis or (2) periodic variations of atmospheric conditions causing differences in the amount of atmospheric refraction. To determine whether the earth's axis of rotation really undergoes a periodic shift, the International Geodetic Union sent an astronomical expedition to Honolulu, which is nearly 180° west of Berlin, for the purpose of making a twelve months' series of observations of latitude corresponding to twelve months' analogous observations made at Berlin, Prague, and Strasbourg. The observations began on June 1, 1891, and the results were published at Berlin in October, 1892. Lord Kelvin has expressed the opinion that 'they prove beyond all question that between May, 1891, and June, 1892, the latitude of each of the three European observations was at a maximum, and of Honolulu a minimum, in the beginning of October, 1891; that the latitude of the European observations was at a minimum, and of Honolulu a maximum, near the beginning of May, 1892; and that the variation during the year followed, somewhat approximately, simple harmonic law, as if for a period of 385 days, with range of about $\frac{1}{4}$ second above and below the mean latitude in each case. This is just what would result from motion of the north and south polar ends of the earth's instantaneous axis of rotation, in circles on the earth's surface of 7·5 metres [8·2 yards] radius, at the rate of once round in 385 days.'

'Sometime previously it had been found by Mr. S. C. Chandler that the irregular variations of latitude which had been discussed in different observatories during the last fifteen years, seemed to follow a period of about 427 days, instead of the 306 days given by Peter's and Maxwell's dynamical theory, on the supposition of the earth being wholly a rigid body. And now, the German observations, although not giving so long a period as Chandler's, quite confirm the result that whatever approximation to following a period there is in the variations of latitude, it is a period largely exceeding the old estimate of 306 days.' This fact goes against the theory that the earth is an absolutely rigid body, and favours the idea that the earth is an elastic body, and therefore subject to deformations. Professor Newcomb has decisively shown that from 1865 to 1890 the variations of latitude were much less than in 1890, indicating that a sudden augmentation took place.

The cause of the variation of latitude has not yet been decided. M. Tisserand, the Director of the Paris Observatory, has shown that the transportation of a mass of water 0·10m. deep, from latitude 45° N. to 45° S., and covering one-tenth of the earth's surface, may cause the principal axis to move 0″·16.   And Lord Kelvin has remarked, 'When we consider how much water falls on Europe and Asia during a month or two of rainy season, and how many weeks or months must pass before it gets to the sea, and where it has been in the interval, and what has become of the air from which it fell, we need not wonder that the distance of the earth's axis of equilibrium of centrifugal force from the instantaneous axis of rotation, should often vary by five or ten metres in the case of a few weeks or months.'   The facts are summed up as follows:—(1) The latitude of a place decreases and increases by one-quarter of a second of arc in a period of 385 days. (2) The change is due to a real shift in the direction of the earth's axis, and not the result of differences of refraction.   (3) The cause of the shift of the axis of rotation is probably the movement of large quantities of water on the earth's surface.

The discussion of the **lesser variations of geographical features** scarcely falls within the scope of the book.   The actual reconstruction of maps of the past from stratigraphy is a task rather for the advanced student of geology than for the physiographer, demanding as it does a detailed acquaintance with the local variations of the minor geological subdivisions. The reader who may find this line of work attractive, as many do, should consult the works of Mr. Jukes Browne *(Building of the British Isles)*, Professor Hall *(Physical History of the British Isles)*, and Dr. Roberts *(The Earth's History)*.   The general principles however are simple.   Coarser strata indicate the proximity of land, shingly conglomerates a shore line, and those with rounded pebbles a river current.   The fact of a bed growing coarser in any direction indicates that the source of its material probably lay in that direction.   Thus, the gradual transition of the carboniferous limestone to calcareous sandstone as one passes from the northern counties of England into Scotland, marks an approach to a shoreline running across the Scotch midlands.   Overlap is, of course, clear evidence of a subsiding coast.   Thus, the carboniferous limestone overlaps on the Charnwood forest (Archæan) rocks, clearly marking a peninsula or island across middle England during the lower carboniferous period and separated from Scotland by a clear warm sea.   The nature of the organic remains and their condition is also of course evidential.   Corals, for instance, indicate clear and probably also rather warm water.   Red sandstones with few fossils and beds of gypsum and salt mark salt lakes, and Geikie has mapped out a system of lakes which lay in an elevated continental area in these latitudes from the disposition of the Old Red Sandstone rocks. The presence of derived fossils and pebbles is also very important as evidence. Thus we find in the New Red Sandstone on either side of the Pennine hills fragments of the carboniferous limestone which makes up that range. From this it has been inferred that this anticlinal ridge was bent up into a ridge even before the close of the Palæozoic period.

The distribution of animals and plants is an important factor in questions of geographical reconstruction, at least from the Eocene onward. The survival of marsupials in Australia and their entire extinction by the higher mammals in the main continental mass of the old world points to a separation of these land masses dating from before the appearance of the more perfected types. The absence of the forms common to Europe and Africa from the fauna of Madagascar points to the conclusion that Madagascar was cut off from the South African mass before the recent connection of Europe and Africa was established. This recent connection is illustrated by the African character of the pre-glacial fauna of Europe with its lions, bears, apes, monkeys, elephants and rhinoceroses. Such large mammals were driven southward by the glacial cold, and the establishment of the Sahara and Mediterranean barriers cut off their return under returning genial conditions. The extreme antiquity of the isolation of New Zealand is shown by the absence of any native mammals save only one or two species of bats, which may conceivably have crossed wide oceanic intervals with the help of drift.

QUESTIONS ON CHAPTER V.

1. How have the temperature, density and salinity of the deepest parts of the ocean been determined?

2. Give an account of the results which have been obtained by recent researches in connection with the amount and the effects of the great pressure to which the lower layers of the oceanic waters are subjected.

3. What is known concerning the difference of chemical properties of fresh water and the salt water of the ocean?

4. Give an account of the results which have been obtained by recent researches in connection with the representation of deep oceanic deposits among the stratified material of the earth's crust.

5. Describe the results that have recently been obtained in connection with the variation in the latitude of a single place, and the conclusions to which they lead.

6. Give a full account of the abyss deposits.

7. Discuss the probable causes of the oceanic circulation.

8. Give a general account of the data from which past geography may be reconstructed.

9. What are the chief views held with regard to the permanence of great oceanic areas ?

10. Give an account of the results of recent investigations with regard to the composition of sea water.

11. Give an account of recent researches in connection with variations of latitude.

CHAPTER VI.

WIDER VIEWS OF THE EARTH'S STRUCTURE.

SUPERFICIALLY it would appear more methodical if we now discussed such questions as sub-aërial denudation and earth sculpture, but as a matter of fact it will be a more profitable order to place before the reader at this stage certain divergent views relative to the condition of the earth's interior and the earlier stages of its history. These fundamental views profoundly affect the consideration of rock structures, mineral veins, and earthquakes, to which we shall next proceed. Taking for granted a general knowledge of the elementary facts of geology, we subsequently shall go on to the consideration of the condition of the known crust of the earth, as distinguished from the unknown substratum of which we shall have speculated. Finally we shall say a little about the general distribution of life in space and time.

The facts bearing upon the internal temperature of the earth are given at considerable length and very ably by Mr. J. C. Christie in the 'Advanced' book in this series, and to that work we may very conveniently refer the student. Proceeding thence to the wider question, we may state that—

The condition of the interior of the earth has long been the ground of a friendly controversy between the mathematicians on the one hand and geologists on the other. It is to be regretted that some minor controversialists on the geological side have seen fit recently to adopt a less amicable tone. The purely mathematical method commences with general assumptions of the internal condition and proceeding with mathematical rigour to deduce consequences, continually corrects its data by the comparison of these results with actual phenomena. The purely geological starts with isolated phenomena and works back towards the final hypothesis. There is certainly on the part of many geologists an unreasonable prejudice against the physicists, and an altogether unscientific objection to the importation of mathematical methods. In some of the utterances of the advocates of hammer and museum, in this for instance: ' the dry flour of the mathematical mill makes excellent paste for holding together the *disjecta membra* of a popular lecture, but it never seems to have commended itself as digestible pabulum to the taste of the every-day geologist who has his day's work to do,'—we have the very same spirit that would have animated the criticisms of some belated alchemist upon Cavendish, Newton, or Boyle. The true tendency of modern science, however, is to value facts only so far as they support or correct great generalisations.

The original view of geologists presented the earth as a thin crust with a molten interior, the crust being the result of superficial cooling, but this was assailed by W. Hopkins in 1847, who gave as the result of calculations based in precession and nutation, a minimum thickness of 800 miles. His conclusions have been questioned by Delaunay, the effects of rotation being disregarded in his calculations. Lord Kelvin (still better known perhaps as Sir William Thomson) has investigated the question from the point of view of tidal action. A thin crust would, he considers, yield with the deformation of the liquid interior, and so no perceptible tides would occur. He requires a crust of at least 600 miles. The Rev. O. Fisher, however, has reinvestigated the question mathematically, and concludes that on an earth with a liquid interior and solid crust ocean tides would exist, of the same type and about two-fifths of the amount of those upon a solid earth. In this clash of mathematicians the student of physiography may very well be advised to suspend his judgment.

Lord Kelvin, moreover, holds that no possible rigidity of a thin crust could prevent its rupture and collapse into a liquid interior.

were a thin crust ever formed. He advances a view of a
practically solid sphere, and assuming that the earth magma first
solidified towards the surface and became denser and sank inward
on solidification, he pictures that this sphere has an internal mass
of such inwardly collapsed fragments forming a 'honeycombed'
structure, in the interstices of which masses of the fluid magma
are caught.

It must be borne in mind although we talk of solid and liquid in
this question, and although in most statements of this controversy
the phenomena of melting and solidification are tacitly assumed
to hold through the range of all temperatures and pressures, we
know practically nothing of the laws affecting the states of matter
at such extreme temperatures and pressures as must obtain
towards the earth's interior. The melting point of a solid which
expands on melting rises with the pressure, though the relation of
the pressure and melting temperature does not appear to have
been precisely formulated. A mass of rock at a considerable
depth may remain solid for a great length of time, but it may
expand, liquefy, and force its way upward so soon as the superincumbent pressure falls, or laterally when lateral pressure
diminishes. It is not necessary to hypotheticate subterranean
stores of *liquid* lava to supply the great outpourings that have
occurred. However, it has been advanced by critics of Lord
Kelvin that such vast outpourings as are indicated by the north
eastern lava plains of the United States, covering areas of over
100,000 miles, are incompatible with his views, and only explicable
on the supposition of at least a continuous sphere of molten rock
beneath a thin crust.

The oscillations of the earth's crust are also advanced against
the view of terrestrial solidity. The case of the so-called temple
of Jupiter Serapis, at Puzzuoli, has been instanced as strongly
suggestive of the rise and fall of a thin crust above an unquiet sea
of molten rock. This appeals to the imagination more strongly
than to the reason. It is forgotten that a solid mass not homo-
geneous will also undergo changes of form with variations in
temperature.

Then the extreme crumpling of strata in the formation of
terrestrial mountain masses is considered by some to be incompatible with great thickness in the crust, which would, they
assert, if thick, be thrown into great bends of small curvature.
But where vertical as well as horizontal lateral pressure acts upon
superincumbent layers, they are thrown into much finer plications

than would be the case if they were folded by lateral compression only, a consideration which does not seem to have had its due weight in this objection. In this connection the student should also read the views of Professor E. Reyer on folding, given later (page 131).

Opposed to the views of Lord Kelvin we may take those of the Rev. Osmond Fisher. He considers that even on *a priori* grounds, the crust may be regarded as perhaps less than twenty-five miles thick, and that it rests on a hollow or solid sphere of liquid rock, possibly a hollow one with a solid nucleus, certainly of concentric layers of increasing density. In this molten rock he finds it necessary to hypotheticate convection currents, radial and tangential, but in his book, the *Physics of the Earth's Crust*, he fails altogether to account for these currents. He needs these currents to prevent the crust getting thicker by melting off the bottom of that crust as it forms. He has since attempted to show that these currents are to be accounted for in connection with the heating effect of the tidal action within. The crust does not collapse into the substratum on Airy's hypothesis that it is practically in a state of hydrostatic equilibrium. He supposes the sub-oceanic crust to dip less deeply into the liquid substratum than the continental crust at the sea board. The continental crust must be less dense than the sub-oceanic and the substratum again less dense than its overlying crust.

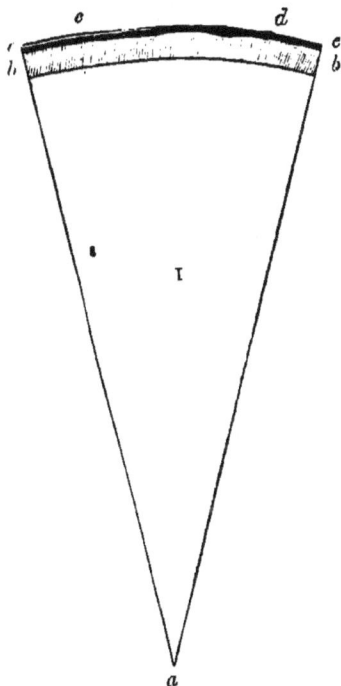

Fig. 20. Segment of Earth according to Fisher's theory. *a* centre, I nucleus, *b* liquid substratum, *c* crust, *d* land, *e* ocean.

The details of his scheme undergo very great alteration in the two editions of his book (1889 and 1891). His views seem on the whole to be far more acceptable to geologists than those of Lord Kelvin.

The chemical as well as the physical composition of the earth's interior is also a purely speculative question. Its specific gravity is too high for it to be a large aerolite and too small for the assumption that it is a siderite. Terrestrial magnetism is of course no evidence of the presence of uncombined iron, since not only are some iron compounds ($Fe_3 O_4$, *e.g.*) magnetic, but a coil of wire through which an electric current is passing may behave exactly like a magnet. This view, that terrestrial currents due to the action of the sun are the causes of the earth's magnetism, was propounded by Grover in 1849, and is now generally adopted. Kant's hypothesis, however, would lead us to suppose that its composition can have no generic difference from that of meteorites, and the general similarity of meteoric and terrestrial minerals certainly favours this view (see *Advanced Physiography*, p. 271, *et seq.*).

Past Chemistry of the Earth.—Some singularly interesting suggestions with regard to high-temperature chemistry formed the bulk of the Presidential Address delivered by Professor Emerson Reynolds to the Chemical Section of the British Association in September, 1893.

As the student is probably aware, as we go upward in the scale of temperature we pass the melting points of an increasing number of bodies. Then we come to their boiling points. At the absolute zero of temperature—in the practical absence of energy, that is—we may conceive of all substances as solid. We may also imagine that a temperature in the upward scale may be attained at which all substances will be in the condition of gas. Moreover, gaseous *compounds* at high temperatures *dissociate*, that is, become resolved into their constituents. Ammonium chloride, for instance, breaks up into ammonia and hydrochloric acid, and these again at still higher temperatures into nitrogen, hydrogen, and chlorine. In the sun we have apparently only dissociated elements, and possibly those substances which have resisted all attempts at analysis under terrestrial conditions (iron, *e.g.*) may be resolved into simpler constituents.

Professor Emerson Reynolds took his hearers back to the time when our globe had no solid core, when the greater proportion of its mass must have been in the gaseous condition. The temperature must then have been far above the dissociating point of oxides or of carbon compounds.

He then proceeded to point out the many points of similarity between carbon and silicon, the agreement of physical properties of the uncombined elements, and the parallelism of their chief compounds. Both elements are tetradic, both are solids, with crystalline, graphitic, and amorphous forms; the hydrogen, sulphur, and halogen compounds are generally parallel, and quite recently nitrogen compounds of Si have been demonstrated equivalent to those of C.

He then went on to suggest that at that time silicon may have played the *rôle* of carbon, and aluminium that of nitrogen, in a great series of compounds comparable to the series of carbon compounds that form the subject matter of the organic chemistry of to-day. The endlessly varied silicates and aluminium silicates that form the bulk of our rocks are, he considered, the oxydised vestiges of that period of chemical activity.

As temperature fell carbon came into play with increasing energy relative to the silicon compounds. The formation of a series of compounds of carbon, nitrogen, and hydrogen, and their subsequent oxydation, with its consequences, must be intimately connected with the phenomena of the beginnings and progress of life on the earth. All the silicon compounds have been oxydised, and the general tendency of vital activity is towards the oxydation of carbon compounds and the formation of carbon-dioxide.

These suggestions, taken together with the results of Professor Dewar's investigations (page 82), will enable the student of physiography to form a general conception of the *history of the world, regarded from the chemical standpoint.* In the past we see a high level of energy—dissociated elements in a gaseous condition. From this state, as energy is radiated as heat into space, we come to the beginning of combination, and as the downward course continues, an increasing number of elements come into play, complex interchanges occur, and more stable and more inert unions become possible. And following the tendency further, we see in the future a continually diminishing violence of reaction, once incompatible compounds lying down side by side ; a great peace, as it were, creeping over the struggle of chemical affinities, until at last, so far as the vision of science can pierce at present, we must come to the absolute zero and the final Death of Matter. This, at least, is what recent chemical enquiry makes the terrestrial prospect appear to be. We may anticipate that the earth and the other constituents of the solar system, their energy radiated away, will roll lifeless through space, until the impact of collision with other 'dark matter' shall disengage fresh energy, liquefy or change to gas all, or some, of the mass, dissociate some or all chemical compounds, and so start over again this cycle of down-ward chemical change.

Determination of the Earth's Density.

(1.) By the 'mountain' method. In this the attraction of a mountain of known weight is compared with the attraction of the earth.

(2.) By the 'torsion-balance' method. In this the disturbance produced upon a pair of small balls by bringing com-paratively larger masses near them is compared with the attraction of the earth, as given by their weight.

(3.) The 'chemical-balance' method. The variation in the weight of a small body produced by bringing a large mass near it is measured and compared with the pull of the earth, as given by the body's weight.

(4.) By 'gravitational' methods. The attraction of gravity is measured at different distances from the earth's centre, by means of pendulum observations.

We will describe these in order. By the law of gravitation the attraction of the earth, that is, the force of gravity (g) at any point, varies directly as the mass (M) of the earth, and inversely as the square of the distance (D) from the centre. In like manner the attraction (a) exerted by a mountain varies directly as the mass (m) of the mountain, and inversely as the square of the distance (d) from the centre of mass. Hence we get the following proportion :—

$$g \; : \; a \; :: \; \frac{M}{D^2} \; : \; \frac{m}{d^2}$$

Therefore $\dfrac{g}{a} = \dfrac{Md^2}{mD^2}$

That is, $\dfrac{M}{m} = \dfrac{gD^2}{ad^2}$

To find the ratio of a to g, the difference of *geographical* latitude between two points on opposite sides of a regular shaped mountain is calculated from the figure of the earth. A plumb-line is then successively suspended at each of these points. The difference between its two directions gives the *astronomical* difference of latitude. If the mountain did not exist, the astronomical difference of latitude would be the same as the geographical difference. On account of its attraction, however, the plumb-line does not take up the direction of a radius of the earth, but makes an angle with it, and this angle is equal to $\frac{a}{g}$, or the tangent of the angle of deflection. The reciprocal $\frac{g}{a}$ is therefore equal to the cotangent of the same angle. The distance D is known, for it is the radius of the earth; d is calculated after the mountain has been measured as accurately as possible; the determination of m is a matter of some uncertainty. It is found by obtaining samples of the rocks of which the mountain is made and determining their density. An average density is then calculated, which, when multiplied by the mountain's size, gives the required mass. All the data are thus obtained from which M can be computed.

In the torsion balance method, two small balls are connected by a fine rod, suspended from its centre by means of a fine wire. Two large masses are then brought into action and tend to pull the balls out of their original position. The angle through which the rod turns is measured and corresponds to the angle of deflection of the plumb-line in the 'mountain' method. The effect (a) produced by each disturbing mass (m) at a distance (d) from the small balls is given by the equation

$$a = \frac{m}{d^2}$$

The pull (g) of the earth upon the small balls is measured by their weight and is given by the equation

$$g = \frac{M}{D^2}$$

where M is the mass, and D the radius, of the earth.

From this we get the same proportion as before, viz.

$$g : a :: \frac{M}{D^2} : \frac{m}{d^2}$$

$$\text{Or,} \quad \frac{M}{m} = \frac{gD^2}{ad^2}$$

It has been said, g is given by the weight of each small ball, D is the earth's radius and is therefore known, a and d are experimentally determined, so is m. The number of times that the earth's mass exceeds the disturbing mass can therefore be calculated. Let x represent the density of the earth and y the density of the disturbing mass, then the following proportion holds good

$$m \; : \; M \; :: \; y \; : \; x$$

from which the density of the earth is found.   This also applies to the previous case.

In the first experiments with the 'torsion balance' method, balls of lead were used to disturb the suspended balls. In order to increase the accuracy of the determination of the angle of deflection, a small mirror was afterwards fixed upon the torsion thread, and caused to reflect a beam of light. By the laws of reflection, the reflected beam turns through double the angle through which the mirror is twisted, and thus ensures increased delicacy of observation.   Air currents also vitiate the result.   To obviate this and other influences, Cornu fixed four globes at the corners of an oblong.   These could be filled with mercury when desired.   The small balls and their connecting beam were suspended between the globes, and the whole arrangement was placed in an air-tight chamber. . To perform an experiment, two globes, diagonally opposite, are filled with mercury and the deflection of the reflected beam observed.   They are then emptied ; the two other globes are filled and the deflection in the opposite direction is measured. From these measures, the earth's density is calculated in the manner above described.

Professor C. V. Boys has devised an entirely new arrangement based on the 'torsion balance' method, in which he uses a very short beam suspended by a fine quartz fibre, and cylindrical weights for disturbing masses. His results, however, have not yet been published.

In the chemical balance method, the theory is practically the same as that of the torsion balance. A body is hung from the beam of a very delicate balance and is found to have a certain weight, which represents, of course, the pull of the earth upon it. A disturbing mass is brought under the body; the effect is observed, and used to determine the earth's density as before. The formula actually employed by Professor Poynting, who has used this method, is

$$\Delta = \tfrac{3}{4} \frac{g}{G \pi R}$$

where Δ represents the earth's density ; g, the acceleration of gravity; G, the gravitational constant, that is, the attraction in dynes which a mass of one gram exerts upon another placed one centimetre from it in air ; π, the ratio of circumference to diameter (3·1416); and R, the radius of the earth.

The gravitational attraction (g) at any point on the earth's surface is equal to M divided by D^2. In descending towards the earth's centre a shell of matter is left behind, and the effect is the same as if it were entirely removed. On this account, therefore, gravitational attraction would gradually diminish from the surface to the centre if the earth were of uniform density. But, on the other hand, the attraction must be increased by the change of position, for it is inversely proportional to the square of the distance from the centre. Hence, at a depth (d) from the surface, there is a shell of matter (m) above, and the gravitational attraction (G) is given by the equation

$$G = \frac{M - m}{(D - d)^2}$$

The intensity of gravity is measured at the earth's surface and in a mine of known depth by means of pendulum observations. The following proportion is thus obtained:—

$$g : G :: \frac{M}{D^2} : \frac{M - m}{D - d)^2}$$

Whence, $\quad \dfrac{M}{M - m} = \dfrac{g\, D^2}{G\, (D - d)^2}$

The chief source of error in this determination is that involved in finding the mass of the shell. Samples of the rocks of which it is made up have their density determined as in the mountain method, and from the results an average density and a probable mass is calculated. The mass of the earth is then found from the equation, and knowing the size of our planet, its density is obtained, for density $\times$ volume $=$ mass.

A similar method to the preceding one is to observe the intensity of gravity at the earth's surface and on the summit of a mountain. At the surface $g = \dfrac{M}{D^2}.$ At the mountain top the distance from the centre of the earth is greater. But the mountain has an extra mass (m). Calling (d) the distance of the pendulum at the summit from the centre of mass of the mountain, the intensity of gravity $G = \dfrac{M}{(D + d)^2} + \dfrac{m}{d^2}.$ Hence we get the following proportion:—

$$g : G :: \frac{M}{D^2} : \left(\frac{M}{(D + d)^2} + \frac{m}{d^2} \right)$$

and from this M can be obtained as in the previous case.

K

Hundreds of experiments have been made by these different methods. The smallest value obtained is 4·7 by the 'mountain,' the largest 6·5 by the 'mine' method. Taking the whole of the results of different observers, the average density of the earth is found to be 5·58.

The origin of mountain ranges has been especially studied by Mr. Mellard Reade, who has propounded a theory of the origin of mountain ranges by Sedimentary Loading and Cumulative Recurrent Expansion. There is a general consensus to regard the great ridges and troughs of the earth's crust as huge wrinkles due to its secular cooling and contraction. However, the attempts to analyse the causes which determine elevation or depression over any region are still at the stage of suggestion and controversy. All great mountain ranges, Mr. Reade points out in expounding his views, are composed of great thicknesses of sedimentary and volcanic deposits, and in some cases we find an aggregate thickness of eight or ten miles of strata, showing, as he considers, that they must have been deposited in steadily subsiding troughs, the original bottoms of which must have finally sunk to twice the depth of the existing ocean. As the mass gradually sinks it will become heated, and this will finally result in an upward expansion and increased lateral tension due to its efforts to also expand laterally. Moreover, as the mass sinks, it becomes a surface of a continually smaller sphere which adds to the lateral tension, and the nett result is upheaval in the centre of the erstwhile sinking area. To him it is objected by the Rev. O. Fisher that the heat which passes into the sediment must be withdrawn from the immediately adjoining parts, which will contract therefore as much as the sedimentary mass expands, and the algebraic sum of these two processes will be not expansion and upheaval but that at most they will balance each other. Mr. Reade, however, has pointed out that the heat is not obtained at the expense of the rocks below those deposited. These rocks to begin with are at the temperature of the surface, but the accumulating sediment acts like a blanket and catches the heat that would otherwise radiate from the region under consideration into space.

Mr. R. D. Oldham has summarised the known facts of the formation of the Himalaya range. His conclusions will serve as an illustration of some of the main views of mountain origin. Nummulites (a foraminiferal organism of Eocene age) occur in the Upper Indus valley at heights of 19,000 and 20,000 feet, pointing clearly to the fact that at that relatively recent epoch the Himalaya region was more or less deeply submarine. On lower slopes the Eocene rocks lie conformably on Mesozoic formations which at that time therefore had not been upheaved. A huge reversed fault separates the older Tertiary and Mesozoic rocks of the Himalayas from the Siwalik series (younger Tertiary), and nowhere on the Himalaya side of the fault are the latter found or any indication that they were ever deposited there. They attain a total thickness of 20,000 feet, showing that they must have been formed during a period of continuous local subsidence. They consist of conglomerate, sandstone and clays, and resemble the present day deposits, which are gravel, sand, and silt, in the fact that they are coarser towards the mountain and thin out and become finer southwardly. A fault similar to that between the older Himalaya rocks and the Siwaliks separates the Siwaliks from recent masses.

Mr. Oldham considers that the first elevation of the Himalaya mass was due to lateral compression and contortion. The elevation once commenced was aided by the continuous denudation of the elevated masses rendering that portion of the crust lighter. The great mass of the denuded material, however, would be deposited near the area of denudation and upheaval, so soon as with the diminishing slope the denuding waters lost their portative power. Hence while the Himalaya region continued to be lightened of material and to rise, the crust in the Siwalik region adjacent became continually heavier with deposits and sank. The strain between the two was ultimately relieved by the formation of the great fault. Later, however, for some reason which is not very apparent, the Siwalik region became involved in the process of upheaval. The region of deposition and depression was thus shifted southward and just as the strain between older Himalaya masses and Siwaliks was relieved by faulting, so was that between the Siwaliks and the recent masses.

A very remarkable and suggestive paper on this subject, by Professor Reyer, appeared in *Nature* (July, 1892). He points out that the contraction of a sphere is not sufficient to account for the deformations the earth's surface exhibits. To these have been added (as we have already stated) the disturbance of the mechanical and thermal equilibrium due to sedimentation and erosion. Then there is loading and disburthening through erosive exchanges. His paper amounts to a proposal to add to these agents another; the crumpling due to lateral shifting of quasi-plastic strata upon an inclined base.

He objects to the view that the earth's crust is in a state of 'magma static' equilibrium, such as the thin crust view of the Rev. O. Fisher involves, the facts that:

(1.) Subsidence does not continue so long as sedimentation goes on.

(2.) That sinking is often considerable where sedimentary loading is slight. (Pacific Ocean, *e.g.*)

(3.) That in many cases enormous loading leads to no depression as in the case of volcanic chains growing up on a highland.

His view of the importance of lateral slipping as a cause of foliation of strata and upheaval, he has illustrated by experiments

Fig. 21.

Fig. 22.

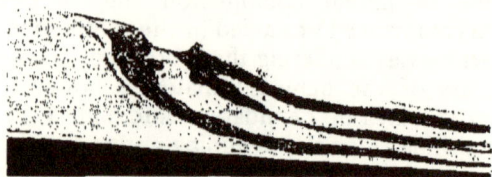

Fig. 23.

Sections of Professor Reyer's model strata to show successive crumpling and distortion as the base (black) is inclined.

made upon the behaviour of model strata of variously coloured clay and plaster upon tilted planes. He has produced in that way remarkably good imitations of folded, contorted, and elevated strata. He points out that the general view that foliation is due

to secular contraction and is a general crumpling of the whole
crust, is disproved by the fact advanced by Brögger, that the
Silurian rocks of Christiana are intensely folded and repose on an
undisturbed base.   The Jurassic strata of the Weser chain are
also folded upon unfolded older strata.   Such cases are only
explicable on the supposition that the strata in question are formed
by a gliding motion over the rocks below, comparable to that of
his model strata.

His views will be best illustrated by the description of the
genesis of an imaginary mountain system.

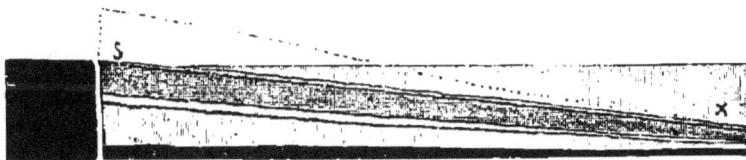

Fig. 24.

In Fig. 24, S is a shore line, S to X are the surface lines of a
mass of sediment derived from the land (black) to the left and
naturally thicker nearest the land.   The shading above X is sea.

This sedimentary accumulation near the land raises the
geoisothermals there, the mass below expands and the sediments
are raised above water.

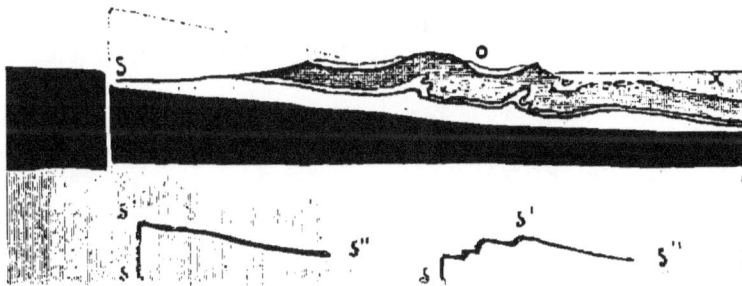

Fig. 25.

In Fig. 25 we see, as a consequence of this, a crumpling and
upheaval of the sediments at O, and fresh sediment accumulating
at X, which is still submerged.

The continual erosion of the original land to the left, leads to cooling and subsidence. As a consequence of this, faulting

Fig. 26.

occurs. Through the faults and fissures eruptive material may escape to form an igneous background to the crumpled masses to the right.

Since Professor Reyer's paper appeared it has received considerable support. Among others, Mr. Mellard Reade has written his appreciation. He considers the theory of 'gliding' particularly applicable to foot hills such as occur at the base of the Canadian Rockies and Himalayas, but he objects to its being made an explanation of *all* folding.

Closely connected with these questions is the discussion of the **origin of crystalline schists** which was exhaustively treated at the Congrès Geologique International in 1888, but without any concord resulting. Two main lines of theory are apparent ; one regarding the Archæan crystalline schists or their lower portion at least, as fundamentally different in nature from superior stratified rocks, and another which finds in them only an exemplification of extreme metamorphism, the causes being still effective. Hunt has subdivided these views as follows :—

The *endoplutonic* view regards the rocks in question as the solidified primitive crust of the previously liquid earth. (Kayser, Roth.)

The *exoplutonic* view is that the crystalline schists are entirely igneous in origin. (Hicks.)

The *chaotic hypothesis* is that they are sediments from a primordial 'chaotic liquid,' probably hot, deposited upon some unknown substratum. (Gümbel.)

The *metamorphic theory* is that they are quite ordinary sediments altered by heat, pressure, and infiltrating water. (Lapworth.)

Hunt's '*crenitic*' *theory* supposes an extensive change of material in a solidifying magma by thermal springs.

Kayser argues that the first solidified crust would probably be acid in composition since it would be the specifically lighter superficial layer of the liquid earth which would first solidify.   He is of opinion that its composition would be the average composition of the subsequently formed sedimentary rocks, which would, in view of the large amount of uncombined silica, be certainly acid.   It would also be invariably the lowest rock, since it has no base, and it would be universally distributed.   He regards these conditions as fulfilled by the lower (Laurentian) gneiss.   His views are strongly tinctured by the continental view that the Palæozoic volcanic rocks are relatively more acid than those of later origin, a view to which British geologists do not subscribe.

A succession of three chief types of rock is a very constant character of the Archæans.   At the base is the gneiss, which may roughly be described as a schistose granite having as constituents quartz, felspar, and either hornblende, muscovite mica, or biotite mica.   *Augite may also occur in gneiss.*   Then follows mica schist, alternating layers of mica and quartz with accessory minerals. Phyllite is a finer mica schist, with a greater proportion of mica, and having as a consequence a silky lustre.   Towards the upper parts quartzites, sandstones, and conglomerates occur, and throughout the series crystalline limestone and graphite.   The gneissic rocks are the Laurentian of Logan and Sterry Hunt, the newer rocks the Huronian.   Abnormal mineral associations occur in the gneissic rocks as compared with undoubted igneous ones.   Silicates of alumina occur side by side with alkaline silicates, and quartz with augite.   These minerals in solution at the same time would react on one another.   The micas of granite have crystalline contours and so probably crystallised out before the other constituents of that rock ; the micas of gneisses on the other hand are interlaced and continuous, and probably therefore formed subsequently to the quartz and felspar.   However gneisses were formed, these conditions are incompatible with the supposition that they were solidified direct in their present state from a molten magma.   Michel Lévy concludes that they were identical in composition with sedimentary schists, modified by contact metamorphism, and injected with eruptive rocks, ' the veritable primitive substratum of the earth's crust,' but remelted and repeatedly re-arranged.

J. Roth has advanced a view that the entire Archæan series was the primitive crust, but this is scarcely tenable in face of the certainly sedimentary constituents of the upper series.

A second view is that all the Archæan rocks are metamorphosed ordinary sediments. There is strong evidence that schists are derivable by lateral compression from sedimentary rocks in the Bergen schists of Silurian age bearing trilobite and graptolite remains, the Carboniferous schists of the Alps, the Jurassic schists of the same region, and the Californian schists of Cretaceous age. This however, as Kayser points out, fails to account for the regular decrease upward in the coarseness of the crystalline constituents. Moreover, the Archæans must have been metamorphosed before the Cambrian period, since Archæan pebbles of gneiss and mica schist occur in Cambrian conglomerates. Dr. Hicks, in agreement with Professor Bonny, denies the practical identity in structure of the newer and older crystalline schists, and asserts that large areas of crystalline schists are always pre-Cambrian. Gneisses he regards as igneous, and the mica and chlorite schists as volcanic ashes and muds. Professor Lapworth and others argue from the theory of organic evolution that the varied fauna of the oldest Palæozoic rocks must have had a long ancestry, which is incompatible with the supposition that the rocks immediately below the Cambrian are the primitive crust. He looks for the demonstration of many Archæan systems only locally altered as investigation proceeds. (See also p. 153.)

Gümbel is one of those who have suggested that the Archæan rocks are chemical precipitates from the superheated primordial ocean, but in this case the minerals should be arranged in the order of their solubility, and this is not the case.

The age of the Earth, like the question of the condition of its interior, is one over which the physicist and geologist appear to be hopelessly at loggerheads. The original calculations of Lord Kelvin, based upon the rate of cooling of the earth, have been recently reconsidered with improved data, chiefly supplied by the experiments of Dr. Carl Barus on the effects of heat and pressure on various rocks, by Mr. Clarence King (*American Journal of Science*, ser. 3, xlv.), and he concludes that the age of the earth probably does not exceed twenty-four million of years. Geologists, on the other hand, basing their calculations on the observed rate of deposition of sediment, require a vastly greater period for the accumulation of the geological series. Sir. A. Geikie requires 100 to 680 million years. Biologists also, on anything but exact grounds however, consider the time given quite inadequate for the evolution of the various forms of life

from a common ancestor such as their science assumes.   The
estimate of Dr. A. R. Wallace is, however, not so widely incompatible with that of the mathematicians.   Assuming a mean rate
of denudation of one foot in 3,000 years over the exposed land,
and the thickness of stratified rocks as 177,200 feet, he arrives at
a total of twenty-eight million years as approximately the earth's
age.

QUESTIONS ON CHAPTER VI.

1. What are geoisotherms? State what you know concerning the position of
these within the earth's crust.

2. What are the chief sources of error in observations directed to the
determination of the rate of increase of temperature within the earth's crust ?

3. Give an account of the results which have been obtained by recent
researches in connection with the nature of those great earth movements which
take place during the formation of mountain chains.

4. Compare the density of the earth as a whole with that of the material
forming its crust, and state the hypotheses which have been advanced to
account for the difference which exists.

5. State the chief arguments for and against the hypothesis that the earth
is solid to its centre.

6. How has the density of the earth been determined ?

7. Give the chief views held by geologists respecting the origin of the
Archæan crystalline schists.

8. Describe the structure of any great mountain range with which you are
acquainted, and explain how far this structure throws light upon the origin of
the chain in question.

9. Give an account of the present state of our knowledge respecting the
probable age of the earth.

10. How has the crumpling of rocks upon an undisturbed base been
explained.

CHAPTER VII.

THE EARTH'S CRUST.

The chief facts in the accumulation of terrestrial materials and the sculpture of the land surface of the earth will already be familiar to the student. We may, however, here call attention to two matters that have recently excited attention. One of these is the process of **subterranean erosion,** the other and much more important one is the discussion of the nature of glacial deposits, and the quantitative estimate of the erosive powers of ice.

It has been pointed out recently that in addition to that solvent action of percolating water which is so conspicuous in limestone districts, there may be a considerable mechanical erosion over an impermeable, underlying a permeable stratum. Streams of water running over a clay and through the substance of a limestone to their points of emergence as streams, must exert an erosive action similar to, if not so powerful as that which they would have under the open sky. Mr. Rutley has also recently published his view that limestones undergo a gradual process of solution. An abstract of his paper, having the title of the ' Gradual Disappearance of Limestones,' is in the *Quarterly Journal of the Geological Society* for August, 1893.

With regard to **glacial erosion** the theory of Sir A. Ramsay that many lake basins were literally *excavated* entirely by glacial action has been greatly questioned. Sir Archibald Geikie, in his *Scenery of Scotland,* finds deep lakes and glacial striæ and detritus so frequently associated that he says 'one cannot but feel, though the problem is not wholly solved, that rock basins are inseparably interwoven with the glaciation of the regions in which they occur.' ' It is generally admitted,' says another modern writer, ' that in confined valleys glaciers are powerful detritive agents and quite capable of sweeping out loose detritus or decomposed rock which lies in their path. In every country the surface must have presented many lake basins of pre-glacial age partly filled with sediments, and these at least may have been swept out by the ice masses.'

Professor Bonney has recently summarised the negative evidence
in the question of ' do glaciers excavate ?' before the Royal Geographical Society. He appeals to such deep lake basins as the
Jordan valley and the Caspian sea, which have never been at a
sufficiently low temperature to undergo glaciation, as a consideration that such lakes may be formed without ice action. He also
contends that the Alpine valleys should be typical instances of
land ice action, and yet ' this is the sum of their evidence,' he says,
' toothed prominences have been broken or rubbed away, the rough
places have been made smooth, the rugged hill has been reduced
to rounded slopes of rock (like the backs of plunging dolphins).
But the crag remains a crag, the buttress a buttress, and the hill
a hill.' . . . . In short,—' Valleys appear to be much older
than the Ice Age and to have been but little modified during
the period of maximum extension of the glaciers.'

Dr. Wright has shown that the great Muir glacier in Alaska
covers wide stretches of undisturbed gravel in which upright tree
stems remain, and similar evidence from the Glacier des Bois and
the Argentière glacier, according to Professor Bonney, shows the
inability of these masses to plough up a boulder bed even with the
advantage of a change of level.

Professor Bonney also contends, with great apparent reason,
that the lakes of Constance, Geneva, Como and Maggiore, instanced as typical glacier-like basins, are comparatively near the
inferior extension of the great ice sheet, and would therefore be
for only a relatively short period under glacial action. Moreover,
he contends that their configuration, especially the radiation of the
outline of Lucerne, Lugano and Como, is not what we should
expect from this theory of their origin.

Professor James Geikie has asserted that the maximum excavation of fjords and lochs is where the maximum pressure of the ice
would be, but this is traversed in several instances by Professor
Bonney.

Professor Bonney's view is that rock basins are ordinary valleys,
formed by subaërial erosion, affected by differential earth movements.
This is strongly supported by the fact that the Iroquois beach of
Lake Ontario is 600 feet higher at the north-eastern than at the
western end, which of course points directly to such extensive local
differences of movement as his view requires.

Of course lakes are frequently formed by the damming up of
valleys by moraines. That is quite another matter to the question
of the excavation of lake basins.

Leaving the consideration of the superficial modification and accumulation of rock materials we may proceed to consider recent work bearing upon the **subsequent internal modification of rocks.** Such investigations fall under one of two heads ; experimental work upon the fusion of rocks, their subsequent cooling, and their behaviour under great pressures ; and speculative work based upon the macroscopic and microscopic study of rock structures. Most important under the former head are the attempts of MM. Fouqué and Lévy to reproduce rocks artificially, Mr. Rutley's experiments upon the devitrification of glasses, and M. Daubrée's on the effects upon rocks of high pressure gas.

The artificial manufacture of rocks dates from as early a period as the time of De Saussuree (1779). In the more recent experiments of MM. Fouqué and Lévy (1882) either the chemical elements or the mineralogical constituents of various rocks were fused in platinum crucibles in a gas furnace. These investigators succeeded in reproducing augite, felspars, leucite, nepheline and garnets. Their rocks since they were cooled under ordinary atmospheric pressure and comparatively rapidly were, as one would expect, of the lava type. They made basalts, phinolites and tephrites, imitation meteorites and also dolerites, and (?) diabases. These artificial rocks resembled to the slightest detail natural rocks. In eruptive rocks, for instance, crystals of hornblende frequently occur with augite and magnetite around them. This has been perfectly imitated by dropping crystals into molten rock, and in this way this structure has been explained.

Mr. Rutley has been studying for some years the process of devitrification in glasses. By heating glasses for long periods he has been able to trace very completely the slow aggregation of mineral particles to form minute crystals. He distinguishes four stages in the process of crystalline growth in a glassy matrix. We may have —

(1.) A *Primitive stage*; in which there occur in the glassy magma little dots alone *(globulites)* or in strings like strings of pearls *(margarites)*, small rods *(longulites)*, which may be simple and as thick at the end as at the middle, or club shape *(clavulite)* or pointed *(spiculite)*.

(2.) A *Spherulitic stage* in which these minute crystalline fibres are gathered together to form round, brush-shaped and bow-like aggregates.

(3.) The *Scopulite* is rod-like in the centre, terminating at either end in a brush of fibres, and thence we pass to

(4.) The *Chiasmolitic stage* in which such rods form regular scaffoldings *(arculites)* disc-like masses, and small crenulated crystalline forms.

M. Daubrée has recently been conducting a series of important investigations upon the action of gas jets at enormous pressures, and there can be no doubt that his experiments throw a very great amount of light upon many geological problems. He has had the use of the appliances for testing explosives in the Laboratorie Centrale des Poudres et Saltpêtres. The high pressure gas is obtained by the explosion of gun-cotton and dynamite, by an electric spark in a very thick steel cylinder, and the aperture of the cylinder contains the rock to be experimented upon. The pressures ranged from 1,100 to 2,400 atmospheres. The rock masses were usually previously divided by a diametrical plane and usually also they were perforated by a fine canal of about 1 millimetre diameter. The most striking and significant results were as follows : —

(1.) Original channels are enlarged (one through granite of 1·2 millimetre to 11·0 millimetres) and fresh ones formed along lines of weakness.

(2.) The substance thus blown out was partly in the form of a fine dust of mineral constituents, and partly in round brownish particles which had apparently been remelted.

(3.) Channels thus formed had their walls striated and polished either with parallel or radiating lines. This polishing is produced by the gas jet itself and *not* by particles driven along by the gas jet. It is of a character suggestive of glacial striations.

(4.) In addition to this erosion, the rocks were generally broken up and crushed by the shock of the explosion, but under the enormous pressure they had reconsolidated in a way that was reminiscent of the regelation of ice.

(5.) So great was this plasticity that a Carrara marble cylinder was considerably flattened and so moulded to the apparatus as to record the slightest striations.

(6.) Substance blown through the orifices noted above collected in caps and blebs suggestive of volcanic scoriæ cones.

The chief applications of these experiments are as follows : — In the first place, they seem to explain the hitherto quite problematical 'diamond pipes' of South Africa. The diamond deposits there occur in cylindrical cavities of unknown depth, perforating both eruptive and sedimentary rocks and filled with decomposed rock. Their diameter varies from 20 to 450 metres. They occur along straight lines and their walls are smooth and finely striated. They terminate above in little heads of rock, *kopjes.* In their occurring along a straight line they so far resemble Daubrée's channels which were determined by the perforation on the diametrical plane of the rock specimen. The striation of the walls and the filling of rotten rock is also in entire agreement in the two cases. Hitherto no views have been advanced to account for these diamond pipes.

The linear arrangement of volcanoes is possibly of a similar nature. What are called ' craters of explosion,' are particularly similar to Daubrées mammillæ. In Velay, near Confoleus, there occurs such a crater, excavated in granite and having its cone formed of granitic fragments. In such cases and possibly in the case of the diamond pipes, the explosive gas is H_2O. M. Daubrée has produced on a smaller scale and to a less extent, similar results to those he got with dynamite gas, with steam.

The character of the débris suggests that what is called ' cosmic dust ' and which is found in pelagic deposits and elsewhere, may be to some extent the result of terrestrial eruptions (Note (2.) especially).

Igneous Breccias may in some cases have been formed in situ.

What are called *slickensides* are striated appearances found on the faces of faults and having a strong resemblance to glacial striations. It has hitherto been supposed that this appearance was due to the friction of the opposed surfaces of the fault when the fracture occurred. However, it may be objected that a single slipping of the surfaces is inadequate to explain the polish, and it is highly probable that the experiments of M. Daubrée will efface this older view.

We may refer the reader to the *Elementary Physiography* for a general account of rock and mineral examination and the classification and formation of rocks and rock structures.

We may perhaps supplement the account of the classification of rocks in the elementary work by the following, based mainly upon their mineralogical and microscopic characters* :—

A. Sedimentary Rocks.

- (1.) *Of mechanical origin or chemically deposited.*

---

* For the verification of numerous points, Cole's *Aids to Practical Geology* has been referred to here. It is a book absolutely necessary to all advanced students of geology.

Sands.—The basis of these is of course quartz. The quartz grains may be simply rounded or coated over with hydrated iron oxide, calcite, or other matter. In the case of '*crystalline sands*' the grains under the microscope show clear crystalline angles, due to a resumption of the process of crystallisation after the accumulation of the sand, the original round nucleus being distinguishable. The most rounded grains characterise wind-blown sands, the buoyant force of water preventing so much attrition as is possible under the wind. Besides crystalline silica, flint may occur in sands, and felspar, mica, tourmaline, zircon, rutile, and other minerals are fairly common.

Grits and Sandstones display the constituent grains of a sand cemented by quartz, amorphous silica, hydrated iron compounds, calcite, &c.

In Gravels and Conglomerates, the pebbles may be rounded lumps of rocks of extremely diversified kinds.

Clays consist essentially of the flat crystalline flakes of kaolin (hydrated aluminium silicate). With this may be more or less quartz (which when abundant makes the clay a *loam*), carbonate of lime (which similarly abundant makes the clay a *marl*), hydrated iron oxides (in yellow and brown clays), finely divided iron pyrites (in blue clays), mica flakes, gypsum, and organic matter.

Shale is a more consolidated clay with a distincter lamination in the bedding planes. Specks of graphite are a very constant feature in sections.

Stalactites, Stalagmites, and Travertine are locally deposited masses of carbonate of lime.

Siliceous Sinter is a similar local chemical deposit of silica from hot springs (in New Zealand and volcanic districts elsewhere).

Gypsum and Rock Salt are chemical precipitates of sulphate of lime typically deposited by the evaporation of inland drainage receptacles.

Subsequent to their accumulation rocks may be extremely modified by local accumulations of matter (segregation). Thus, nodular masses of calcium carbonate accumulate in calcareous clays to form *Concretionary Limestones*. Similarly carbonate of iron grows round centres of precipitation in clays to form *Clay Ironstone*. *Ironstones* generally are sedimentary rocks, calcareous, sandy, or argillaceous, largely impregnated or partially replaced by segregated iron carbonate.

Flint and Chert are segregations of hydrated silica occurring frequently in limestone and in sandy rocks. Possibly the siliceous material has been derived from the adjacent rock by the solution of radiolaria, diatoms and other siliceous organisms originally mingled with the deposit.

Concretionary phosphates also occur.

(2.) *Mainly or entirely of material of organic origin accumulated through a greater or less amount of mechanical agency.*

(*a*) *Limestones.*—Hand specimens, or if the elements are finer, sections, of these, show their constitution from organic remains. They may be sub-divided into—

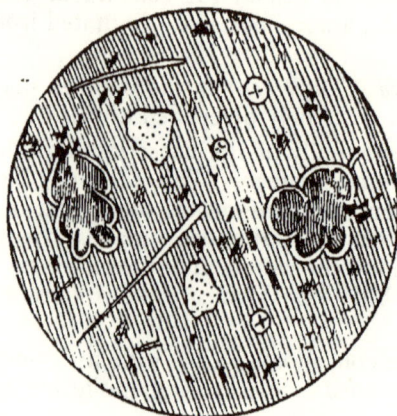

Fig. 27.   Section of Chalk.

Shelly Limestones.— Microscopic examination enables us to determine the constitution of the finer varieties of these sometimes very exactly. The structure of echinoderm skeletons is peculiar, and seems to identify their minutest remains. Mr. Sorby has shown that the so-called 'Stonesfield slate' is a fine limestone largely composed of flakes of oyster and brachiopod shells.

Coral Limestones.

Nullipore Limestones of the remains of calcareous algæ.

Pisolitic and Oolitic Limestones, consisting of small round grains formed by the accumulation of fine calcareous mud around broken fragments of shell rolled to and fro by the waves.

Dolomitic Limestones are cavernous limestones in which the $CaCO_3$ is largely replaced by $MgCO_3$.

Limestones under stress of earth movement may be converted into *calc schists*. (See below Crystalline Limestones.)

(*b*) Plant remains give rise to *peat, lignite, brown coal, common coal and anthracite*. Microscopic examination of coal itself reveals little structure except in the case of what are called *Spore coals*. Sections of these give a brownish black mass with more or less flattened hollow ellipsoidal or circular bodies of a golden red colour. These are the spores of the cryptogamic vegetation which gave rise to the coal deposit. The sections of fossil woods by which our knowledge of palæobotany has been considerably extended, are cut not from coal but from plant fossils found in the adjacent beds.

B. Rocks of Igneous origin.

(*a*) *Holocrystalline Rocks, i.e.*, rocks consisting entirely of fairly developed crystals. They are of deep-seated origin and their crystals are large because the process of cooling has been relatively slow (hundreds of years). The chief types are—

Granite. *Free silica (i.e., quartz), orthoclastic* felspar, and either white mica, ferruginous mica (*biotite*), hornblende or any two or all three of these.

Quartz Diorite. *Free silica (i.e., quartz), plagioclase* felspar, biotite, hornblende.

No other typical holocrystalline rocks have free quartz.

Syenite is a hornblendic granite without quartz. With augite taking the place of the hornblende it is *augite syenite*, with mica, *minette*.

Diorite is similarly related to quartz diorite. But with augite in the place of hornblende it is *Gabbro*. Usually the augite of Gabbro is schillerised (diallage). Where mica replaces the hornblende it is mica-diorite. A finer gabbro is a *Dolerite*.

None of the above rocks contain olivine. Where the basic olivine occurs, we have *Olivine Diorite* and *Olivine Gabbro*. No such thing as olivine syenite occurs. Olivine and quartz, we may further note, are incompatible as rock-forming minerals, since the acidity of the latter would be neutralised by the basicity of the former.

Troctolite is olivine and plagioclase felspar.

The 'ultra-basic' rocks (= the Peridotites), *Dunite, Saxonite, Picrite*, and *Lherzolite*, are rocks containing olivine alone, and with enstatite, enstatite and augite, and augite respectively.

Felspars in many of the above series of rocks may be replaced partially or entirely by the minerals nepheline or leucite ; then we have *Nepheline-Syenite* (from syenite) and *Nepheline Diorite* (from the diorite group), where the replacement is partial, and Nephelinite and Leucitite where the replacement is complete. With olivine, the two latter form Nepheline Olivine Gabbro, and Leucite Olivine Gabbro.

(*b*) *Finer Crystalline Rocks* formed more superficially.

Quartz Felsite corresponds to granite, *Quartz Aphanite* to *Quartz Diorite. Aphanite* and *Dolerite* to the diorite-gabbro rocks and '*Finer Peridotite*' to the 'ultra-basic' group of dunite, saxonite, picrite, &c.

(*c*) Lava rocks containing or not containing crystals, but always some uncrystallized ground substance or *glass*, since the cooling has been relatively rapid (a few years). The more acid ones are commonly more glassy than the more basic.

Rhyolite answers to granite.

Rhyolitic Trachyte to quartz diorite.

Trachyte to syenite.

Andesite to diorite.

Basalt to gabbro.

Olivine Basalt to olivine gabbro and diorite.

Limburgite to the peridotites.

Phonolite to nepheline syenite.

Tephrite, Nepheline Basalt, and *Leucite Basalt* to nepheline-diorite, &c.

Certain rocks named for structural peculiarities cannot be classified in the above scheme, such are lava froth or *pumice, volcanic ash* and the like.

The more basic rocks are particularly liable to alteration through hydration. Altered diorites and gabbros are *Diabases*, altered basalts are *Melaphyres.* The general result of the hydration of the peridotites is *Serpentine Rock.*

C. Metamorphic Rocks.

(a) Rocks affected locally by contact metamorphism.

Spotted Slate containing aggregates of minute crystal 'embryos' (Cole) of biotite, garnet, etc.

Slates with new minerals, pyrites, chiastolite, various micas, garnets, &c.

Baked Shales, hardened into a porcelain-like consistency and with secondary minerals.

Locally crystallized Limestones.

(b) Rocks modified over wide areas.

Crystalline Limestones, through which run at times serpentine. A final stage of the limestone rocks named above, or as a result of the extreme modification of igneous rocks, rich in lime, with a schistose structure, we have calc-schist.

Quartzites, quartz grains of sandstones, cemented by silica. Often with mica. Under the microscope many quartzites show compressed and crushed grains. They graduate into 'flaggy gneisses' of quartz and mica.

Slate.

Schists. Rocks with the minerals in foliæ which may be greatly crumpled. There are few large crystals and they are fissile. Sub-divisions determined by the prevalent mineral are mica-schist, chlorite-schist, serpentine-schist, talc-schist, hornblende-schist and quartz-schist.

Gneisses are more coarsely crystalline and with a well-marked felspathic element. They are less readily split than the schists. Commonly the gneisses are acid in composition and typically they contain quartz, felspar, and mica, as do granite or quartz diorite. Basic gneisses do not occur, there is nothing to correspond with serpentine schists. What are called 'gabbro gneisses' appear to be merely local conditions of igneous masses.

Within the last twenty years the **microscopical examination of rocks** has become greatly extended, and the results obtained have very profoundly affected many geological questions. The method was first applied by Nicol in 1827 to the examination of fossil wood and was subsequently extended to rocks and minerals by Sorby in 1856. A special form of microscope is used bearing a removable substage Nicol's prism, and another capable of being inserted into or withdrawn from the microscope tube.

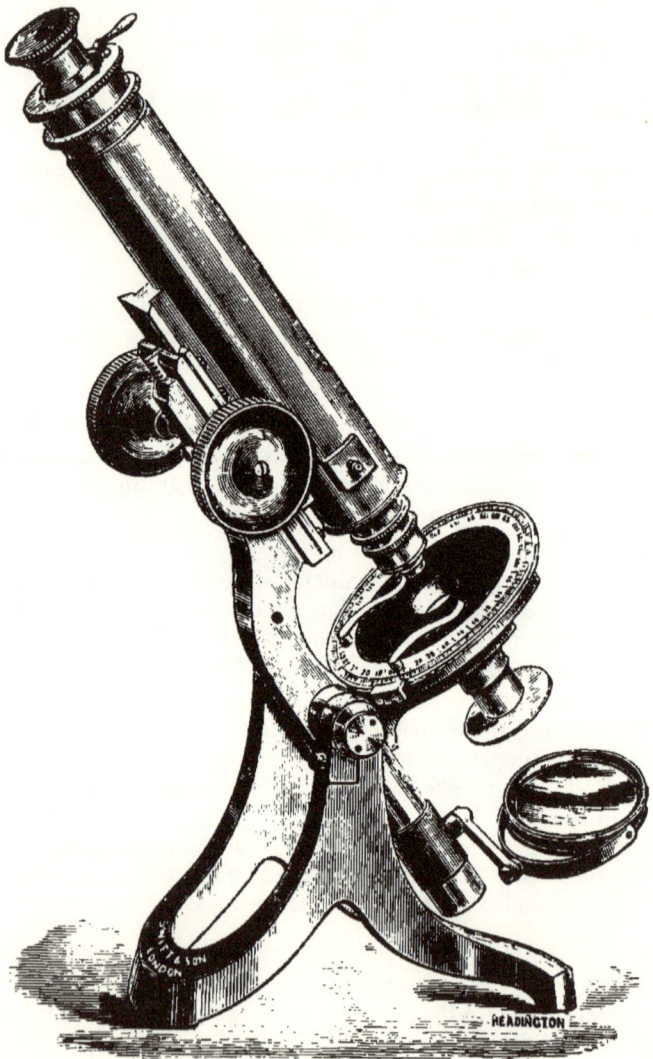

Fig. 28.

We give a figure of a new type (Fig. 29) of petrological micro-scope produced by Messrs. Swift, of Tottenham Court Road, and also their common form, used by the students at the Royal College

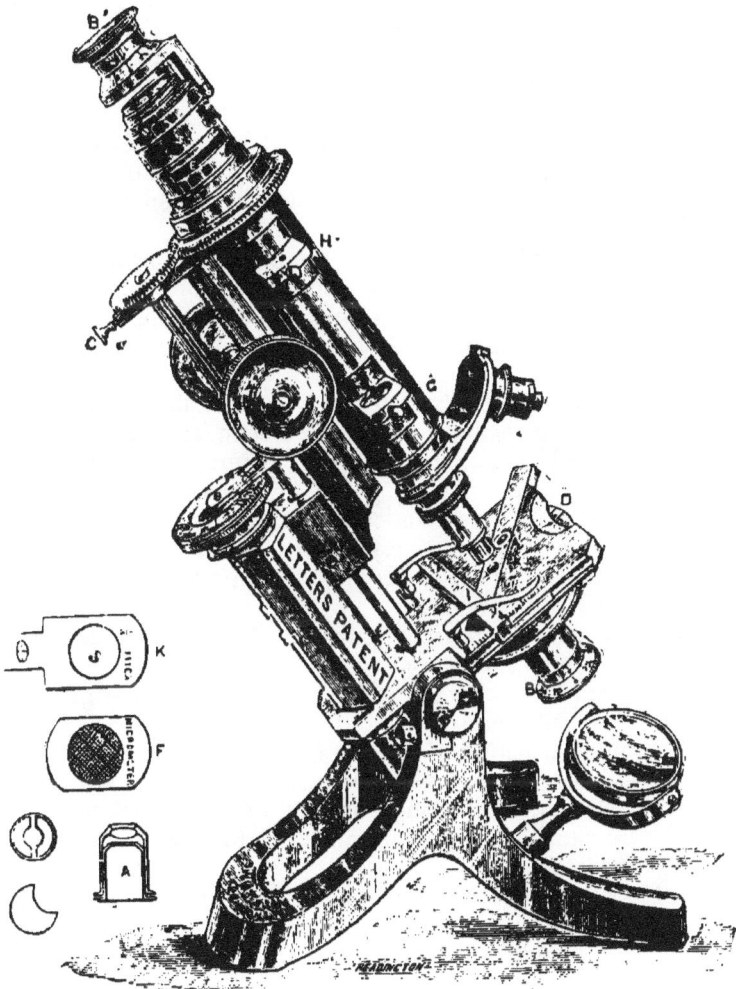

Fig. 29.

of Science (Fig. 28). The former instrument has been designed by Mr. Allen B. Dick. At B, under the stage which carries the slide, is the polariser, a Nicol's prism, which may be rotated either alone or simultaneously with the analyser by means of toothed wheels. It has a catch which fixes it in a 'crossed' position relatively to the analyser. The 'diagonals' of both are then parallel to the cross wires of the eye-piece. They can all be turned simultaneously by the wheels either crossed, parallel, or with their 'diagonals' inclined 45° to the cross wires.

At B' is another Nicol's prism, the analyser. It is capable of the same movements as the polariser. It has a steady pin which moves in a slot to allow of the analyser being drawn out to enable the observer to focus the cross wires when his vision is abnormal. For this purpose the eye-lens has a long screw. The analyser may be removed altogether, as may the polariser, so that the slide under consideration may be examined by ordinary light, in polarised light from the polariser only, or between both Nicols.

In common practice minerals are identified by such characters as the following :—

(1.) The crystalline character. Non-crystalline bodies and those belonging to the cubic system are black between crossed Nicols, and have their natural colours when the Nicols are parallel, giving in intermediate positions merely intermediate states of illumination ; but *most* sections of bodies crystallised in all other crystalline systems than the cubic, give interference colours between the Nicol's prisms.

(2.) These colours are frequently to a certain extent diagnotic, olivine and most augite give extremely rich tints; quartz, characteristic bright colours; nepheline, pale grey and drab tints and so forth.

(3.) Twining structure of crystals is brought out by the Nicols, adjacent turns having complementary tints. Orthoclase (felspar), for instance, is told from plagioclase by the brilliant banded appearance due to repeated twining of the latter.

(4.) Pleochroism is an alteration of colouring, as the polariser is rotated when the analyser is withdrawn. In this way hornblende, which is strongly pleochroic, is distinguished from augite.

(5.) The tendency of crystals to crack along definite lines (cleavage) if judiciously observed is often valuable, *e.g.*, between hornblende, augite (both of which have characteristic cleavages), and olivine.

(6.) The presence or absence of alteration products. Felspar, for instance, is often clouded through alteration, quartz is invariably very clear, olivine almost invariably accompanied by secondary serpentine and magnetite.

(7.) The refractive index relative to Canada balsam shown by the darkness or otherwise of the outlines and cracks, and the appearance of mottling on the surface of the mineral (in olivine, *e.g.*) in ordinary light.

(8.) Inclusions, their character and distribution. Those of quartz are characteristic, and much more so the dark masses arranged parallel to the outline of the crystal in nosean, and in the form of a cross in chiastolite. So also are those of diallage described below under '*inclusions.*'

(9.) Crystalline form. Sphene, for instance, often gives wedge-like sections, nepheline, various sections of the simple hexagonal prism, magnetite, octahedral outlines, and mica a characteristic frayed outline.

The discussion of the more refined methods of identification involving the use of the quartz plate, the separate examination of the interference figures seen by a special adjustment of isolated crystals in converging light, and so forth, belong rather to the sciences of optics and mineralogy than to physiography.

Sections of sufficiently indurated rocks are prepared for examination by first cutting slices by means of a lapidary's wheel, polishing one side by means of emery powder and rouge, attaching firmly to a glass slip by means of Canada balsam, and then rubbing down the other side until the section is transparent. This applies more particularly to the igneous and harder sedimentary rocks. Loose sands and clays are examined by mounting fragments on glass, and it is also sometimes convenient to crush more compact rocks and mount the powder in the same way.

The particular value of this method of investigation is in the refinements it has introduced into the study of crystalline rocks. But it has also clarified many a doubtful matter with regard to sedimentary structure. We need only remind the reader, for instance, of the light thrown upon the nature of chalk and oolitic grains. However, its most valuable application is undoubtedly in the new views it has given of metamorphic processes. Thus the process of heat metamorphism of clay, as an intrusive mass of igneous rock is approached, may be demonstrated, and the gradual development of audalusite, quartz, mica, staurolite, and garnet, traced. There is also a tendency in minerals having an identical

chemical composition to gravitate from a less stable to a more
stable form.   Calcite replaces aragonite, hornblende, augite, and
this is shown by crystals having the outer form of the less, and

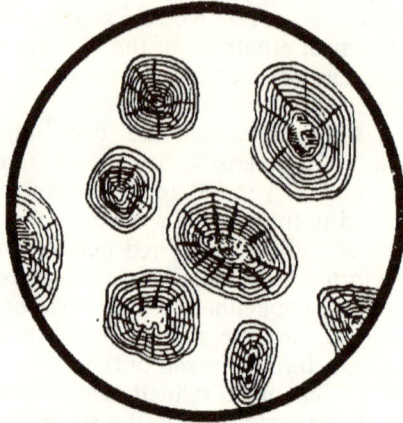

Fig. 30.   Oolitic Grains.

the minute peculiarities of the more stable form.   Felspar under
metamorphic agencies breaks up partly or altogether into quartz
and white mica, silicates may be replaced by carbonates, and
granite become converted into tourmaline rock.   The formation
of such minerals as epidote, chlorite, and serpentine, and the
relation of these to the original minerals of the rock is another
matter of importance.   By the comparison of selected sections,
while keeping in view macroscopic structures, a great amount of
light has been thrown upon endless geological problems.
    We may give one or two particular instances of the application
of this method of research.   Thus the successive stages in the
crushing of a granite, and the establishment of what is called by
Professor Lapworth, *mylonitic* structure, have been described by
Professor Bonney.   Sometimes the rock is merely crushed, sometimes there is a shearing or dragging movement in the mass
tending towards a banded structure.   In the simple crushing the
quartz is broken up, the felspar powdered, the mica flakes bent,
twisted, and torn.   With the microscope we can trace every stage
of this from the minutest deformation of the original crystals.
Moreover, water is intimately mixed into the masticated rock, and
chemical action is set up as a consequence of the intimate contact.

Black mica (biotite) is decomposed, felspar dust gives rise to white mica and chalcedonic quartz. Parallelism of the minerals and even foliation may be set up, and a coarse granite may be seen to be converted according to its more or less acid composition, into a quartzose or micaceous schist. With such evidence as this the probability of gneisses and crystalline schists being metamorphosed igneous rocks becomes an absolute certainty.

The secondary nature of some, if not all, hornblende, as a sequel to its isomer augite, is indubitable. Some petrologists regard hornblende as invariably evidential of the action of pressure but with this Professor Bonney does not agree. Hornblende may occur in three main series of forms in rocks.

(1.) With its own proper crystalline outlines.

(2.) As *uralite*, which is hornblende with the outline of augite, that is, the outline of the augite crystal has not been altered, but its substance has passed into the condition of hornblende, and has the characteristic internal cleavages and pleochroism of the latter mineral.

(3.) As blades and needles, even as *actinolite* (fibrous hornblende).

Professor Bonney considers that hornblende of the third type is the only form developed in connection with excessive pressures.

He has applied his researches particularly to the problem of the Archæan crystalline schists. He considers the conditions under which these were formed no longer act in the earth's crust, and that they are not identical in structure with 'crystalline schists' of later date. He concludes that the schists and gneisses have been formed in one of three different ways, which are stated by him as follows:—' (1) Some have once been molten, but have become solid under *rather exceptional circumstances, probably having lost heat slowly, and having continued to move very gradually during the process of consolidation.* (2) Others have been produced by the thorough alteration of sedimentary materials, in which *a high temperature has been maintained for a long time* in the presence of water and under considerable pressure. (3) Others, again, have been the result of great pressure which has acted on rocks already crystalline, and has produced mineral changes, sometimes to the complete obliteration of the original structure.'

In another place we discuss this question of the origin of the crystalline schists more fully (p. 136). (1) above, refers, it would seem, particularly to the lower Archæans which Professor Bonney

thinks may have been formed before any oceanic water was condensed, and when the pressure even at the surface of the solidifying
crust due to the vast atmosphere of steam and air may have been
300 or more times the atmospheric pressure of to-day. (2) refers
to the later Archæans which solidified when the downward rise
in the internal temperature was enormously more rapid than it
is now.

Another interesting line of microscopic petrology is the investigation of **crystalline inclusions.**   These are foreign substances
caught up by a mineral during crystallisation and enclosed
in its substance.   There are four classes of inclusion; (1) gaseous;
(2) liquid ; (3) glassy, and (4) stony.

Gas inclusions, appear as a rule to contain air or carbon dioxide.

Liquid inclusions usually do not completely fill the cavities
enclosing them.   In some cases the liquid is water, and the small

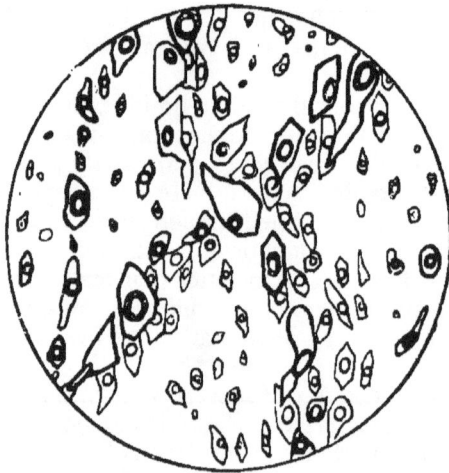

Fig. 31.   Liquid Inclusions.

cubic crystals of common salt have in some cases been observed in
this.   In the case of some of the liquid inclusions, however, heating
up to 30·92, the critical point of CO_2, causes a disappearance of
the liquid, the cavity becoming uniformly filled with gas.
Evidently the liquid here is carbonic anhydride condensed by
pressure.

Glassy inclusions can be distinguished from liquid since the bubble of air or gas in this case is incapable of moving about. They are of course portions of the original magma of the rock.

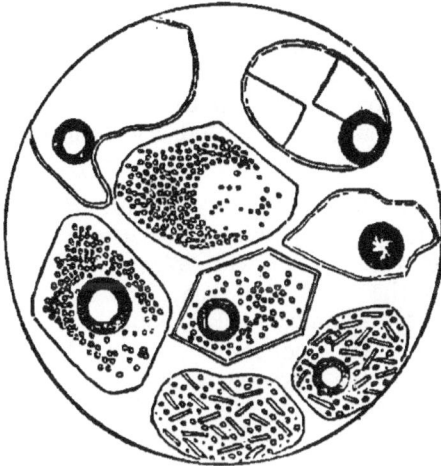

Fig. 32. Stony Inclusions.

Stony inclusions differ from the glassy in the fact of their being devitrified.

In many cases inclusions are arranged in planes corresponding

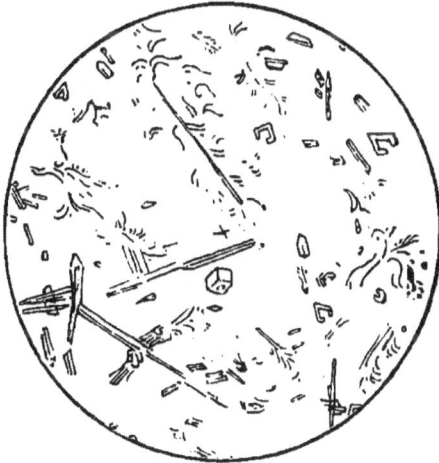

Fig. 33. Microliths.

to plans of crystalline form in the including mineral. This is the case in the schiller structure (schillerisation), of diallage, labradorite and hypersthene, and it has been made the subject of exhaustive enquiry by Professor Judd. The inclusions consist chiefly of hydrated oxides, but really no satisfactory conclusions have been discovered respecting their causation. The water and gas of inclusions so very characteristic of quartz, may have been dissolved in the magma from which the igneous rocks containing them solidified, just as CO_2 may be dissolved in water.

Passing from this matter of rock structure to the phenomena of **hypogene action,** we may notice that latterly there has been but little work done in connection with volcanic phenomena to which we need refer in this book. We may, however, add to what the student will already be familiar with in connection with deep seated disturbances, a few facts concerning the methods of seismology and an indication of the tendency of the conclusions of investigators in this department.

Seismography, the earthquake record, originated in that quivering country, Japan. The first instruments were probably those of Chōkô, dating from the year A.D. 132. Since then, many forms of observational instruments have been designed in Japan, and so far from seismology being a Western science, the Western nations have adopted, or at most re-invented, many Oriental appliances. However, all native Japanese contrivances record from fixed points, which gives precise indications of the direction and amplitude of shocks but not of their duration or intervals such as is obtained by tracings on a moving surface.

The general idea of most modern seismographic apparatus is the same. It is that familiar friend of the physiography student, Foucault's pendulum, in a new character. The plane of vibration of a finely adjusted pendulum is independent of terrestrial movements, and while it continues it defines a constant direction in space. Hence any movement of the earth, local or as a whole, must be movement relative to this absolute datum. The seismograph determines, as nearly as possible, qualitatively and quantitatively what this relative movement is.

The following are some of the chief types of seismographic apparatus, exhibited as typical at the Chicago exhibition. We quote the account of Professor J. Milne given in *Nature*, Feb., 1893.

A. Recording Instruments.

(1.) *Seismograph writing on a glass disc.*—In this we have horizontal pendulums writing the earth's motion as two rectangular components on the surface of a smoked glass plate. The vertical motion is given by a vertical spring lever seismograph. The rate at which the plate revolves is accurately marked by an electrical time ticker. The movements of the latter are governed by a pendulum swinging across and making contacts with a small vessel of mercury. The revolving plate is kept in motion by clockwork, which is set in motion by an electric seismoscope. (See No. 8.)

(2.) *Seismograph writing on a drum.*—In this instrument the record is written on a band of paper, the diagram being less difficult to interpret because it is written to the right and left of a straight line and not round a circle.

(3.) *Seismograph writing on a band of paper.*—In this instrument not only is the diagram written along a straight line but it is written with pencil,—the trouble of handling smoked paper being therefore avoided. When the earthquake ceases, the drum ceases to revolve, but if a second or third earthquake should occur, it is again set in motion. By this instrument a series of earthquakes may be recorded, the resetting of the instrument being automatic.

(4.) *Seismograph without multiplying levers.*—This instrument is intended to record large motions, the horizontal levers not being prolonged beyond the steady points to multiply the motion. For large earthquakes, when the ground is thrown into wave-like undulations, special instruments which measure tilting are employed.

(5.) *Duplex pendulum Seismograph.*—In this case a steady point is obtained by controlling the motion of an ordinary pendulum with an inverted pendulum. The record consists of a series of superimposed curves written on a smoked glass plate.

(6.) *Mantelpiece Seismometer.*—This is intended for the use of those who simply wish to know the direction and extent of motion as recorded at their own house. It is a form of duplex pendulum, and it gives absolute measurements for small displacements.

(7.) *Tromometer.*—This is one form of an instrument which is used to record movements which are common to all countries, called earth tremors. Every five minutes, by clockwork contacts and an induction coil, sparks are discharged from the end of the long pointer to perforate the bands of paper which are slowly moving across the brass table. If the pointer is at rest, then a series of holes are made following each other in a straight line, but if it is moving, the bands of paper are perforated in all directions round what would be the normal line of perforations.

The earth movements which cause these disturbances are apparently long surface undulations of the earth's crust, in form not unlike the swell upon the ocean.

A more satisfactory method of recording these motions, which has been used for the last two years, is by a continuous photograph of a ray of light reflected from a small mirror attached to a small but extremely light horizontal pendulum.

B. Instruments which release the recorders.

Electrical contact makers.—These instruments are delicate seismoscopes, which on the slightest disturbance close an electric circuit, which, actuating electric magnets, set free the machinery driving the recording surfaces on which diagrams are written.

C. Time record.

Clock.—At the time of an earthquake the dial of this clock will oscillate quickly back and forth and receive on its surface three dots from the inkpads on its hour, minute and second hands respectively. It thus records hours, minutes and seconds, without being stopped.

From those fortunately situated investigators of earthquakes in Japan there comes to hand, it is true, an abundance of observations, 121 shocks in one year being recorded by the Gray Milne Seismograph; but up to the present there is nothing very conclusive as to the causes of their frequency in that part of the world. The most disturbed districts are the extremities of peninsulas. No evidence has been brought forward to link these particular earthquake shocks especially with volcanoes, and no law showing any seasonal and diurnal influence has become apparent. Neither has any direct connection of magnetic phenomena with them, been made out. But electrical disturbances accompany earthquakes in many cases and are confined to their region of influence. The air in such cases always becomes electrically negative, and a change of potential of as much as 30 volts may occur.

Of course there can be no doubt that a condition of frequent vibration in the crust is correlated with volcanic conditions, a thing shown very conclusively in the map in the elementary work. Moreover, any internal collapse or rupture such as must occasionally occur in a volcanic region, must necessarily give rise to an earth tremor, just as the explosion of a mine would do. But it does not follow that all earthquakes have an immediate causative connection with volcanic paroxysms. A considerable number of observers assume a connection between the majority of earthquakes and culminating stresses leading to the slipping of faults.

Mr. Davison, for instance, investigated the earthquakes originating in the British area in 1889. In that year five were observed. The centre of the iso-seismic lines is taken as the epicentrum, or point on the earth's surface vertically above the centre of propagation. Of the five earthquakes, one at Edinburgh, in January, was very generally felt, and is connected by him with a large fault along the axis of the Pentlands. The Lancashire earthquake of that year stopped clocks, rang bells, and awakened sleepers. There were sounds, the duration of which was greater on the whole near the line of the Irwell fault. The isoseismals were circular in form, suggesting that the slipping was horizontal. An earthquake occurred in Kintyre in the neighbourhood of no traceable fault. There were two slighter earthquakes in East Cornwall and at Ben Nevis, and moreover two doubtful ones in 1889.

In 1890 there was a very perceptible series of earthquakes in Inverness, one in Kintyre, and one in Yorkshire. The Yorkshire one affected a small area and had a focus near the surface. There is no fault in the neighbourhood of the latter, and Mr. Davison does not ascribe it to fault slipping, but to the collapse of cavities in the magnesian limestone, an explanation also advanced by Lebour for certain Sutherland concussions. For the rest Mr. Davison evidently regards fault slippings, vertical, horizontal, and in intermediate directions, as adequate causes. They illustrate the dying away of the fault-producing forces, and belong to the closing epoch of upheaval, 'the epoch immediately preceding the death of a mountain chain.'

In Mr. Davison's investigation of the Cornwall earthquakes of 1892, he traces them to a fault (not shown in the survey maps) running east and west, and which he regards as one of a series formed during the changes of relative level that resulted in the

Fig. 34.   Earthquake Fissures, Bella near Naples.

depression of the English Channel.   As a matter of fact, however, the Cornwall faults were as a rule north-west to south-east, and not parallel to the channel coast as this suggestion of Mr. Davison's would lead one to infer.

The connection of earthquakes and atmospheric pressure is one that would seem *a priori* undeniable. If there is a condition of stress between two regions of the crust tending to an alteration in relative level, the rupture will be determined, other things being equal, by such a distribution of barometric pressure that a cyclonic area will lie over the region straining upward, and an anticyclone over the downthrow area.

The delicate seismometers now set up in many observatories record earth-movements much too small to be directly observable. By such means the earthquakes are detected at a considerable distance from the places where they are felt. Thus, an earthquake on the Italian Riviera affected the instruments at Greenwich Observatory. By recording the exact times at which the same earthquake visits different places at a known distance apart, the velocity of the wave can be calculated. (Velocity = distance ÷ time.) The velocity with which the waves travel differs according to the character of the rocks traversed. It is about two miles per second in granite, and less than 1,000 feet per second in sand and clay. Mallet found that an earthquake at Calabria in 1857 travelled in different rocks at rates from 650 to 1,000 feet per second.

The chief **magnetic phenomena of the earth's crust** are described in the elementary work. We may add to what is given there, notes of the chief instruments employed in such observations, on the connection of sun-spots and magnetic storms, and on the distribution of terrestrial magnetism.

Magnetic Elements and their Determination.—The magnetic condition of any place is known when the three 'magnetic elements' have been determined. These elements are as follows:—

(1.) Declination, or the angular distance between the position taken by the north pole of a compass needle and the direction of a true north and south line.

(2.) Inclination, or dip, signifies the angle which a perfectly symmetrical needle moving in a vertical plane makes with the horizon.

(3.) Total Force is the intensity of the magnetic action of the earth at the place of observation.

The essential instruments of a magnetic observatory are (*a*) the declinometer, (*b*) the dip circle, and (*c*) the horizontal force

M

d. Beneath the magnet is a tube having a photogra
at one end in the focus of a lens at the other. A s
:ope is arranged to view the scale through the lens, and v
entral division of the scale is in the centre of the field of v
xis of the telescope lies in the axis of the magnet. An in
e sun or a star can be reflected through the scale to
:ope, by means of a small mirror. If the object were on
lian, that is, due south, at the time of observation, the a
gh which the telescope had to be turned to view it cent
gh the photographic scale would be the declination. B
he azimuth can be found when the time of observatic
l, by reference to the *Nautical Almanac*, and from it
nation or variation of the compass can be calculated.
dip circle is shown in Fig. 35. It consists of a very delic:
iced needle contained in a brass box having plane glass e

The box rests on a base thro
which passes three levelling scr
Two small spirit levels indi
whether the base is horizonta
not. Before the needle was r
netised, the axis of rotation pa
through the centre of gravity
there was no tendency to take
any particular position. After r
netisation, however, in these latit:
the north pole of such a ne
takes up a position below the l
zontal, owing to the action of
vertical force of the earth's r
netism. To determine the ang
dip at any place, the base is
levelled. The box containing
needle rotates independently
the base, and it is turned 1
the needle comes to rest v

35. Cassella's Form of
    D:_ C:__l_

om each other.   A slip of brass, capable of being clam
' position, rotates under the box containing the needle,
.ear its end, a knob on the under side.   Suppose the bo>
:urned so that the needle sets vertically.   The slip of b
 moves independently of the box, is rotated until the k
under end fits into one of the four cups.   It is then clam
.t it moves with the box.   When the box has been tui
it the brass slip fits into the next cup, it has been rot
;h a right angle.   The plane of motion of the needle is
 magnetic meridian.   Both the horizontal and vertical c
its of the earth's magnetism therefore act upon it and c.
ake up a position representing the resultant of their ac
livision behind the north pole of the needle is read off
 angle of dip at the place of observation.
ɔ horizontal component of the earth's magnetism canno
nined directly like the declination and inclination.   A ma
:ed east and west of a small compass needle, the ne
akes up a position resulting from the action of the ma
ɪe horizontal pull of the earth.   Let the moment of
et be M and the horizontal force H.   The deflection ξ
tio of H to M, that is, $\dfrac{H}{M}$.   The same magnet is then I

hread and the time required to make a complete oscilla
.   The oscillations depend, among other things, upon
:nt of the magnet and the horizontal force, so the fe
is determined by this means.   Knowing the values of $\dfrac{H}{M}$

 the force H can be found.   It is equal to the square
ɔ product of the two measured quantities, and is expre
ɪes if the centimetre-gramme system of measures is emplo
otal force is equal to the horizontal force divided by
ɔ of the angle of dip.

fferential Magnetic Instruments.—In addition to
ments for determining the absolute values of magi

time. Three instruments are usually employed in the automatic registration of terrestrial magnetism, one to indicate the declination, another to show the horizontal force, and a third to exhibit changes in the vertical force. From this it will be seen that the angle of dip is not usually registered.

The declinometer (Fig. 36) consists of a magnet about five inches long, suspended by a single fine thread, and having a vertical mirror beneath it. The upper half of the mirror is rigidly fixed to the magnet, the lower half is immovable; hence, when an illuminated slit of light has its image thrown upon the mirror, one half is constantly reflected in a fixed direction, the other half is reflected in different directions according to the different positions assumed by the magnet.

The horizontal force magnetometer also consists of a magnet five inches long, and having one half of a small mirror fixed beneath it, while the other half rests on the table (Fig. 37). The magnet is suspended from two fine wires a short distance apart. The arrangement is such that if a rod of brass were hung instead of the magnet, the two wires would lie in a plane at right angles to the magnetic meridian. On account of the horizontal pull which the earth exerts, the magnet is pulled out of the plane in which the wires were originally fixed. The top of the wires remain in this plane, but the bottom is twisted out of the plane as the magnet tries to set itself in the magnetic meridian. The action of the wires tending to pull the magnet into a plane at right angles to the meridian can be considered as constant. Hence, if the horizontal component varies in intensity, the magnet will assume different positions, and the beam of light reflected by the mirror attached to it will vary in direction, as in the case of the declinometer.

The vertical force magnetometer consists of a magnet having a V-shaped ridge, known as a knife edge, crossing it at the centre.

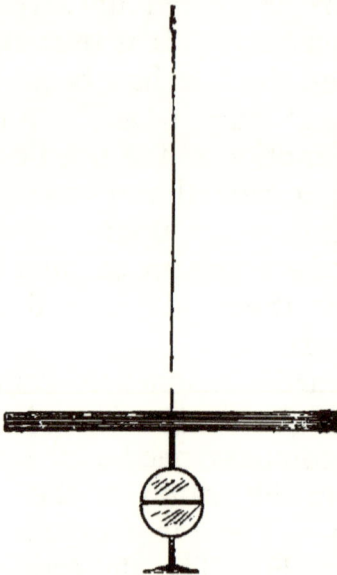

Fig. 36. Differential Declinometer.

The magnet rests with this edge on an agate plane and is balanced so that the south pole dips slightly down. One half of a small mirror is attached to the magnet, as in the two previous cases, the other half remaining fixed. Variations of the earth's vertical force at the place of observation cause the magnet to move up and down about the balancing point, and therefore cause the beam of light reflected from the attached mirror to alter in direction.

The method of recording the changes in the position of the magnets is very simple (Fig. 38). A slit of light impinges upon each mirror and is reflected. The two parts of the mirror produce two images, which, by means of a hemi-cylindrical lens, are reduced to two points of light, and fall as such upon a cylinder covered with sensitised paper. The cylinder rotates once in twenty-four hours. The luminous dot from the fixed half of each mirror traces a line upon the prepared paper, that from the half attached to the magnet is always on the move. The

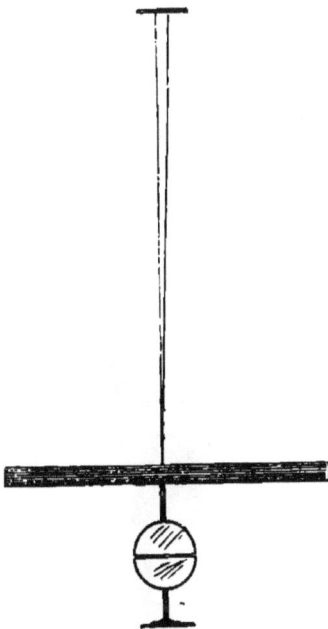

Fig. 37. Differential Horizontal Force Magnetometer.

irregular line thus produced upon the paper is, to a small extent, due to mechanical vibrations of the magnet. But by far the greater part of the variation from the true position is caused by fluctuations of the components of terrestrial magnetism. 'Magnetic storms,' 'diurnal,' and other changes in the magnetic elements are registered automatically in this manner. The instruments for doing this work are thus essential to the equipment of a properly constituted magnetic observatory.

Magnetic Storms and Sun Spots.—Great disturbances upon the sun have on several occasions been accompanied by disturbances of the earth's magnetic elements. The enormous spot-group of February, 1892, is a case in point. On February 13, when the maximum area was reached, magnetic instruments were violently disturbed, and continued to be disturbed all through

Fig. 38.   Arrangement of Continuous Recording Instruments in a Magnetic Observatory.

the day and evening until the following afternoon. The variation in magnetic declination was more than one degree, and the other elements were caused to vary through similar large amounts. The disturbance was greatest during the night of February 13, when an aurora borealis was seen at several places. Fig. 39 is a

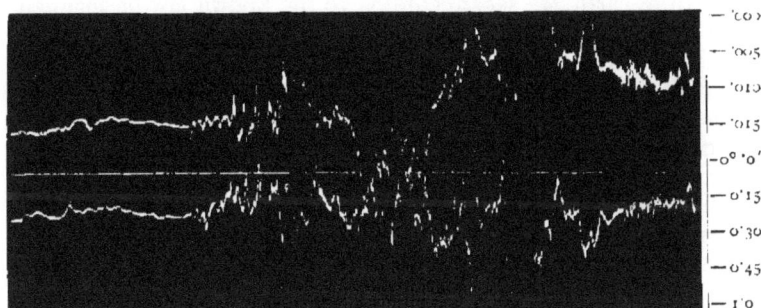

Fig. 39. The Magnetic Storm of February, 1892.

reproduction of traces obtained at Greenwich Observatory from February 11 to February 14. The upper trace represents the photographic record of the movements of the declinometer between those dates, and the lower shows the horizontal force variations. For a short distance from the left-hand side the traces have their normal character, for they are always irregular lines. The true magnetic storm commenced on February 13 at about 5.30 a.m., and the disturbances continued until the afternoon of the following day. There are two or three more similar coincidences, yet it cannot definitely be said that sun-spots and magnetic storms always run together. Large spots have appeared upon the sun without affecting the magnetic elements, and violent magnetic storms have occurred when very few and small spots have been visible.

Lord Kelvin, at the anniversary meeting of the Royal Society in 1892, discussed the probability of a physical connection between sun spots and terrestrial magnetism. In his words :— ' The primary difficulty is to imagine the sun a variable magnet or electro-magnet, powerful enough to produce at the earth's distance changes of magnetic force amounting, in extreme cases, to as much as $\frac{1}{20}$ or $\frac{1}{30}$, frequently, in ordinary magnetic storms, to as much as $\frac{1}{400}$ of the undisturbed terrestrial magnetic force.

'The earth's distance from the sun is 228 times the sun's radius, and the cube of this number is about 12,000,000. Hence, if the sun were, as Gilbert found the earth to be, a globular magnet, and if it were of the same average intensity of magnetisation as the earth, we see, according to the known law of magnetic force at a distance, that the magnetic force due to the sun at the earth's distance from it, in any direction, would be only a twelve-millionth of the actual force of terrestrial magnetisation at any point of the earth's surface in a corresponding position relatively to the magnetic axis. Hence the sun must be a magnet of not much short of 12,000 times the average intensity of the terrestrial magnet (a not absolutely inconceivable supposition) to produce, by direct action simply as a magnet, any disturbance of terrestrial magnetic force sensible to the instruments of our magnetic observatories.'

Lord Kelvin has also investigated the probability and possibility of the sun being an intensely strong magnet, with special reference to magnetic storms. Of a storm of this character, which was registered by the recording instruments in a number of magnetic observatories, he says:—In this eight hours of a not very severe magnetic storm, as much work must have been done by the sun in sending magnetic waves out in all directions through space as he actually does in four months of his regular heat and light. This result, it seems to me, is absolutely conclusive against the supposition that terrestrial magnetic storms are due to magnetic action of the sun ; or to any kind of dynamical action taking place within the sun, or in connection with hurricanes in his atmosphere, or anywhere near the sun outside. It seems as if we may also be forced to conclude that the supposed connection between magnetic storms and sun-spots is unreal, and that the seeming agreement between the periods has been a mere coincidence.'

The Distribution of Terrestrial Magnetism.—By con-necting places on the earth having the same declination at a given epoch, a number of lines, known as lines of equal declination, or *isogonic* lines can be traced upon a map. (Fig. 40.) A similar magnetic map can be constructed by connecting places having equal inclination. The lines thus produced are termed *isoclinic* lines. (Fig. 41.)

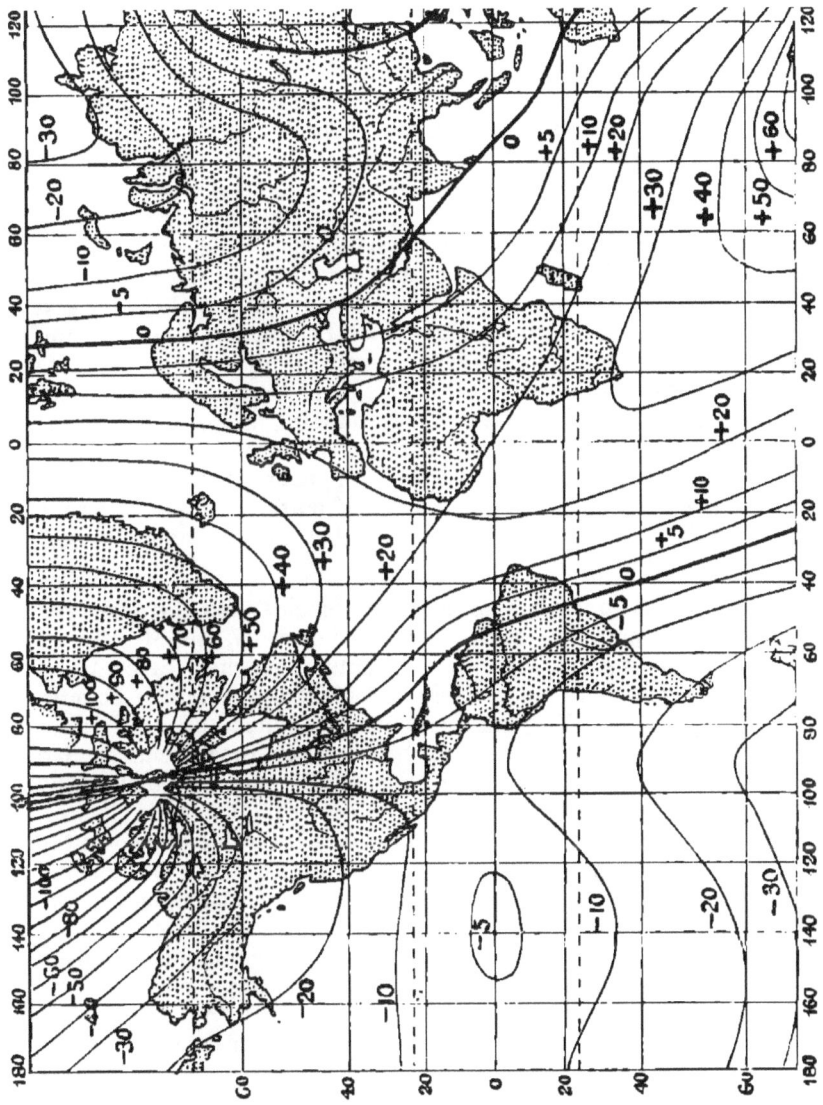

Observations show that the declination is *west* in the eastern portion of North America, in the north-east of Brazil, the Atlantic, Indian Ocean, Africa and Europe. The magnetic needle also points west of true north in Japan and the eastern part of China. On the other hand, the declination needle points *east* of true north in almost the whole of America, in the Pacific Ocean and in Asia. A more or less regular circle, along which there is no declination, limits these zones, and divides the earth into two unequal parts. It passes from the north magnetic pole to the east of Spitzbergen, from North Cape to St. Petersburg, Astrakhan and the Sea of Oman, cuts the western part of Australia and goes to the south magnetic pole. From this point the line rises to South Georgia, passes to the west of Rio Janeiro, across Brazil, to the east of the Antilles, and then to the north magnetic pole through the lake region of the United States. This circle of no declination cuts the earth's equator in longitude $54°$ on the west and longitude $81°$ on the east. It follows from this that the declination is west over $135°$ of longitude and east over $225°$.

The line of no inclination does not coincide with the terrestrial equator. It cuts the equator north of Polynesia, and, if followed towards the west, is found to rise until it reaches the centre of Africa. From this point, it sinks rapidly and cuts the equator near longitude $3°$ W. The descent is continued in a curve towards South America and cuts the coast of Brazil south of San Salvador. After this, the line rises to the starting point. The inclination increases rapidly on both sides of the magnetic equator, and continues to increase until it reaches the value of $90°$ at the magnetic poles. The total magnetic intensity at any place is not observed directly but is deduced from the measurement of the horizontal component and the inclination. It increases from equatorial regions towards the poles. The minimum intensity occurs in the middle of the Atlantic, on the tropic of Capricorn. The foci of greatest intensity do not coincide with the magnetic poles. Two unequal foci exist in each hemisphere. In the northern hemisphere the principal focus occurs near Hudson's Bay, and the secondary focus is found in Siberia. Two similar foci were found by Sir James Ross in the southern hemisphere.

The positions of these foci are not the same as the positions of the magnetic poles—the points at which an inclination needle sets vertically. The latitudes and longitudes of the magnetic poles and the foci of total force or intensity are tabulated below.

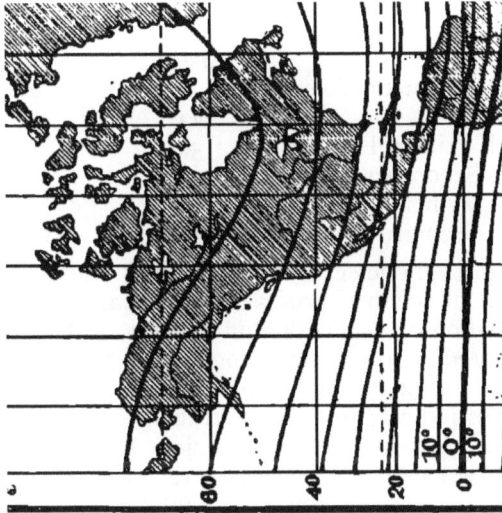

Fig. 41. Isoclinic Lines, or lines connecting plac
inclination is the same. Above the magnetic eq

	Latitude.	Longitude.
North Magnetic Pole	70°	98° W.
South „ „	73	146 E.
North Principal Focus	52	92 W.
„ Secondary „	69	117 E.
South Principal „	66	138 E.
„ Secondary „	52	129 E.

The recent magnetic survey of the British Isles, undertaken by and under the direction of Professors Thorpe and Rücker, involved the determination of declination, inclination, and horizontal force at two hundred stations, and from these data the actual isomagnetic lines were drawn. Their divergence from the mean terrestrial ones indicate the local disturbance of the general effect due to the earth's magnetism as a whole. Where the disturbance of the mean horizontal force ceases to increase and begins to diminish we are passing over a central line of attraction or a 'ridge line,' while where the reverse occurs we are passing the boundary between one centre or axis of attraction and another, a 'valley' line. Where two ridge lines intersect we have a magnetic 'peak.'

No possible causes of these disturbances can be thought of except local earth currents or magnetic rocks below the surface. No connection can be traced with the former, and it is to the latter that we must turn. Messrs. Highfield and Jarratt have experimented with a number of rock specimens from the Malvern Hills, and also from the collection of Professor Judd. The nett result appears to be that the magnetic susceptibility is largely dependent on the amount of magnetite present. Specimens of even the same rock gave very variable results. Various specimens of gabbro from Loch Coruiskh, Skye, give magnetic susceptibility ·00027, ·00049, ·00082, ·00153, ·00284, ·00362, ·00684.

The nett result of the investigation has been the conclusion that the ridge lines and peaks mark near approaches of ferruginous rocks to the surface. We have such a peak near Reading, for instance, and thence three ridges run, one to Chichester, one to Greenwich, and one westward to Gloucester. Professor Rücker, commenting on M. Mascart's survey of France, believes that the 'Palæozoic axis' running southward from Reading can be traced into the heart of France. The ridges of attracting rock hypotheticated by this survey may be made out from the accompanying map.

Captain A. Schück has been conducting a similar survey of the North Sea for some years past (1884—91), and his *Magnetische Beobachtunges auf der Nordsee* gives an account of his methods and results. His investigations have been undertaken in wooden ships as free from iron in their construction as possible, and he has *neglected any disturbance due to his ship*. It is possible that this may considerably invalidate his results, since in ships not specially constructed some magnetic disturbance is almost inevitable. It is proposed to repeat his result at a later date with more favourable appliances.

QUESTIONS ON CHAPTER VII.

1. Describe the nature and origin of spore coals.

2. Give an account of the results which have been obtained by recent researches in connection with the origin of serpentines.

3. Give an account of the results which have been obtained by recent research in connection with the origin of 'crystalline sands.'

4. Describe the composition and characters of the chief types of glassy rocks.

5. Describe the composition and physical characters of the glassy ground-mass, and of the chief crystals which occur in basaltic lavas.

6. Give an account of the results which have been obtained by recent researches in connection with the artificial formation of igneous rocks.

7. How has it been proved that carbonic anhydride is present in a liquid condition in some rocks?

8. Name the chief minerals which occur most commonly as rock consti-tuents, and give the chemical composition of each.

9. Describe the chief methods for determining the height of mountains.

10. Compare the composition of meteorites with that of the earth's crust.

11. How can the rate of movement of earthquake waves be determined by time observations, and what are the chief sources of error in such observations?

12. Describe a magnetic observatory.

(The above questions are from examination papers.)

13. Describe the macroscopic and microscopic characters of gneiss.

14. Give an account of the petrological microscope, and of the chief characteristics by which minerals in rock sections are identified.

15. Discuss the bearings of Daubree's recent experiments upon the influence of high pressure gas in modifying rock structure.

16. What do you know of recent discussions of the influence of glacial action upon scenery?

17. Describe a seismological observatory.

18. How far do you consider volcanic to be connected with earthquake phenomena?

19. What are volcanic dykes?

20. Give a brief account of the results of the recent magnetic survey of the British Isles.

21. Describe the instruments used in a magnetic observatory.

CHAPTER VIII.

GENERAL FACTS OF THE DISTRIBUTION OF LIFE IN TIME AND SPACE.

WE may conclude this book with a brief review of the relations of the phenomena of life to the external universe, so far, that is, as natural science is concerned with life. It is difficult at first to realize how extremely localized and temporary a thing the whole career of life is, compared with the play of lifeless forces. In the infinite extension of space, so far as our evidence goes, life is confihed to the surface of this one small planet of ours; and its extension in time on this earth can cover only a small portion of this planet's career, since the range of its resistance to extremes of temperature is less than that from the boiling point to a few degrees below freezing. Bacteria resting spores, however, can endure a temperature below – 180°, but they manifest no *activity* at such low temperatures although they are not celled. The gaseous, the liquid and the hotter phases of the solid state of this world were lifeless, so also will its latter phases be without life. From the point of view of the stellar astronomer or of the physical and chemical philosopher life is an entirely local thing, an eddy in one small corner of the immense scheme of Being.

The beginning of life is unknown. Lord Kelvin's supposition that the first life was extra terrestrial and arrived upon a meteorite, is open to the serious objection that it could scarcely have survived the fiery rush through the air. Many zoologists consider that there must first have been an aggregation of complex albuminous compounds under conditions of chemical synthesis entirely unlike those that now obtain. It is tolerably certain that the career of life began in the shallow sea, and according to Professor E. Ray Lankester the first forms may have been simple structureless animals, mere streaming pools of living matter, such as are now called 'plasmodia.' From such substance the vast branching trees of the animal and vegetable kingdoms arose.

The earliest traces of life on this earth are radiolarian remains, described by Dr. Charles Barrois, from the **Archæan** rocks of Brittany (*Comptes Rendus*, cxv., 1892, p. 326). They are the only certainly organic remains in these rocks and they are referred by M. Cayeux to the family of the Monosphæridæ. They are accompanied by graphite. That old familiar friend of the student of elementary geology, *Eozoon*, has now been clearly demonstrated to be a purely mineral formation, a precisely similar structure having been shown to exist in ejected blocks from Monte Somma by Doctors Johnston-Lavis and J. W. Gregory.

In the Cambrian rocks, however, the traces of life become fairly abundant. It will perhaps be better adapted to our purposes if instead of our describing the succession of life forms according to their geological systems we follow the zoological and botanical classifications. The whole field of life is thus classified.

A. Organisms with their protoplasmic substance enclosed for all or part of the life duration in **cellulose cell walls.**

= PLANTS.

i. Plants having no woody tissue, and therefore leaving, as a rule, no fossil remains = **Non Vascular Plants**, *e.g.*, mosses, algæ, fungi, and lichens.

ii. Plants having woody tissue = **Vascular Plants.**

(1.) Vascular Plants producing spores which develop into a plant unlike the parent, bearing sexual organs which reproduce the first form again = *Vascular Cryptogams.*

(2.) Vascular Plants in which the spore and sexual stages are abbreviated to form a seed stage = Phanerogams (or seed plants).

The Phanerogams are sub-divided into Gymnosperms (= conifers, cycads, zamias), and Angiosperms, of which the former resemble the vascular cryptogams more closely than the latter.

B. Organisms the protoplasm of which never forms around itself cellulose cell walls.

= ANIMALS.

i. Mere blebs of protoplasm, usually invisible to the naked eye, which usually have no mouth and never any body cavity = **Protozoa,** including foraminifera and radiolaria.

ii. Animals with the body perforated by innumerable *pores* by which nutrient material enters. With skeletons of horn, carbonate of lime or glassy silica = **Porifera** or sponges.

iii. Animals which have usually a simple bag-like form, usually with a mouth in the centre surrounded by arms, *e.g.,* sea anemone and jelly fish. They may secrete a skeleton of horn (graptolites, *e.g.*) or carbonate of lime (corals) = **Cœlenterata.**

iv. Animals consisting as it were of two tubes, a food canal or intestine running through the body and surrounded by the body cavity. Commonly a blood circulation, and invariably nerves and nerve centres = **Cœlomata.**

The cœlomata branch out into a series of great groups, of which the mutual relationships are still largely a matter of conjecture. The chief are :—

(*a*) *Vertebrata ;* cœlomata, with a back-bone, a hollow spinal cord, and either throughout life or in the earlier stages of development with gill slits : fish, dipnoi ('mud fish,' transitional forms between fish and amphibia), amphibia, reptiles, birds, and mammals.

(*b*) *Echinodermata ;* cœlomata, usually with their parts arranged round a centre like the spokes of a wheel round the hub, as in the sea-urchins, star fish, and brittle stars.

(*c*) *Chætopoda ;* long bodied cœlomata, without limbs or a backbone, the body being divided into a succession of ring-like parts repeating each other: earthworms and marine worms.

(*d*) *Arthropoda ;* cœlomata with numerous pairs of jointed limbs. The division includes the extinct trilobites, insects, crustacea (crabs, lobsters, shrimps, &c.), arachnids (scorpions and spiders), millipeds and centipeds.

(*e*) *Mollusca ;* soft, bodied cœlomata, without any division internal or external into rings, without a backbone or true internal skeleton, and usually with an external shell. The shell is bivalve in lamellibranchs (oyster, mussel), generally single and coiled in gasteropods (whelk, snail, winkle), small and light in pteropods when present at all, and in cephalopods (squid, octopus, nautilus), either coiled and chambered, or as a spongy cake or bony rod more or less completely sunk into the animal's body.

(*f*) *Brachiopods ;* superficially resemble *Lamellibranchs* but differ fundamentally in their internal anatomy. They have no blood circulation, the nervous system, the alimentary canal, and the development is entirely dissimilar. Most brachiopods have no anus.

(*g*) *Polyzoa* are small cœlomata of simple structure, usually living together in colonial masses, their bodies being joined. They are represented by the common sea-mats of the shore.

(*h*) Besides these main groups, there are the *Tunicata,* degenerate relations of the vertebrata, the *Rotifers,* possibly the most primitive of the cœlomata, and the parasitic round and flat worms, but these have no skeletal arrangements to become fossilised and have left scarcely any traces on the geological record.

On the evolutionary theory now held by all zoologists and botanists, the first forms of life to appear must have been some such forms as the simpler protozoa. From these were derived, along diverging lines, the existing protozoa, the algæ, and the cœlenterata. The sponges are variously regarded as a special group of cœlenterates, or as separately derived from the protozoa (Saville Kent). The cœlomata are descended by one, or possibly more than one, line from the cœlenterata. The mutual relationships of mollusca, chætopods, arthropods, rotifers, polyzoa and brachiopods are extremely problematical, but there is a general consensus that the chief subdivisions of the vertebrata and the plants are genetically related in the way shown by the accompanying table. If such a descent has really occurred, the protozoa would appear first, then the groups whose names come nearest to *Protozoa* in our table, and so outward to the extremes of the branches.

N

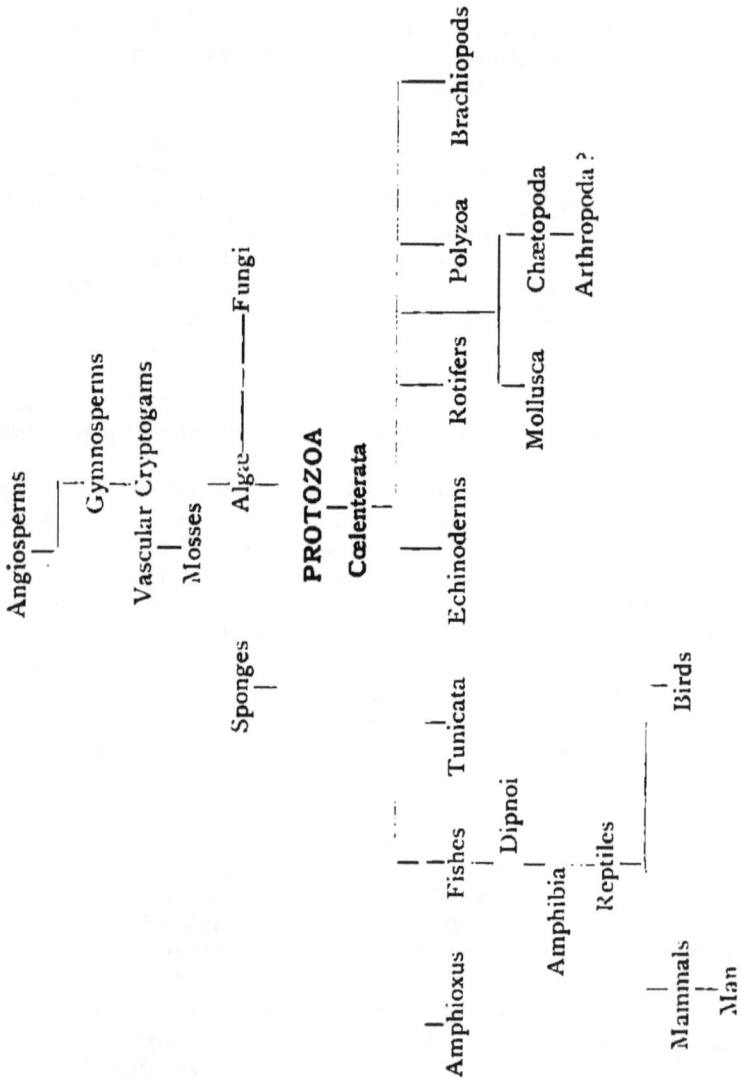

Brachiopods

Polyzoa

Chætopoda

Arthropoda?

Fungi

Gymnosperms

Vascular Cryptogams

Mosses

Algæ

Angiosperms

PROTOZOA

Cœlenterata

Rotifers

Mollusca

Echinoderms

Sponges

Tunicata

Dipnoi

Birds

Amphioxus

Fishes

Amphibia

Reptiles

Mammals

Man

The *Protozoa* we have already noted are the first organisms to appear. This is what the evolutionary hypothesis would lead us to anticipate.

Sponges appear in the lowest fossil beds by their representative *Protospongia fenestrata*.

Most *Algæ* and *Fungi* do not afford hard parts for fossilization. What are called *fucoids* are tracings of extremely doubtful and varied import upon rocks. Some are probably tracks of crustacea and worms; Nathorst has imitated some by trailing medusæ over mud. Some *may* be prints of algæ. Some algæ (*e.g.*, nullipores) secrete a calcareous element in their cell walls. Such may be the problematical *Oldhamia* of the Cambrian.

Vascular Cryptogams, according to Dawson, begin with *Protannularia* of the Ordovician. They are certainly represented in the Devonian. According to the evolutionary supposition they should precede gymnosperms.

Gymnosperms appear in the Devonian with Dadoxylon and are represented by *Dadoxylon, Pinites,* and *Aurucaroxylon* of the Carboniferous.

Angiosperms only certainly appear in the Jurassic. This is in accordance with the fact that anatomically the gymnosperms come nearer to the more primitive vascular cryptogams than do the angiosperms.

Podocarya and other trees annually shedding their leaves, such as are now characteristic of temperate climates, begin only in the Upper Cretaceous.

Cœlenterata are represented in the very lowest Cambrian by certain problematical forms, the *Archæocyathinæ*, which were apparently coral-like organisms.

Echinoderms appear in the Ordovician rocks, among the first being *Echinosphærites*. Of the seven chief divisions of these, two, the Cystidians and Blastoidians, are extinct, and of a third (the Crinoids) only *Comatula* survives.

Mollusca. The first bivalve mollusc is *Cyrtodonta* of the Cambrian; the first gasteropod, *Bellerophon* in the Ordovician. So called pteropods (*Theca* and *Conularia*) appear in the Cambrian, but, according to Pilseneer, these remains are probably not pteropods. *Orthoceras*, the earliest cephalopod, appears in the Upper Cambrian. The appearance of all the distinct groups of mollusca thus early is scarcely what one would expect.

Polyzoa appear in the Ordovician.

Brachiopods are among the very earliest fossils, beginning in the lowest Cambrian with *Lingulla primæva*.

Chætopoda are represented by worm tracks in the early Cambrian.

The *Arthropoda* appear very early in the geological record, earlier than the view that they are more modified *Chætopods* would lead us to expect. Moreover, the great subdivisions are early distinct.

Trilobites are exclusively Palæozoic. *Palæopygæ* appear in the Lower Cambrian, and *Phillipsia* ends the race in the Permian.

True *Crustacea* first appear in the Lingula (Cambrian) rocks with *Hymenocaris*. The modern lobsters only become important in the Mesozoic, and the highly modified crabs are Tertiary.

Arachnids (scorpion and spider group). A great group of water scorpions, the *Eurypterids*, are exclusively Palæozoic. A true scorpion appears in the Silurian (*Palæophonus*), and a spider in the Carboniferous.

Insects (*Palæoblattina*, a near relation of the cockroach) appear in the Upper Silurian of France.

If the *Arthropods* jar a little with the evolutionary view the distribution of the vertebrata in time is in entire agreement therewith. The simpler aquatic forms precede the more complex land inhabitants, and the growth of knowledge completes more and more the links in the chain of their development.

Traces of *Fishes* appear first in the Ludlow bones bed of the Silurian, and their remains are abundant in the Devonian. Some of these latter may be Dipnoi. *Ceratodus* of the Triassic rocks is a still living genus in Queensland rivers. The Teleostel, that section to which most living fish (except the sharks and rays, and the sturgeon) belong, only appear in the Cretaceous.

Amphibia are first represented by the Carboniferous *Labyrinthodonta*.

Reptiles certainly appear in the Permian (*Mesosaurus*).

Birds are first represented by the archaic long tailed *Archæopteryx* of the Jurassic rocks, but they are only abundant in Tertiary rocks.

Mammals appear in the Triassic rocks with the small, half reptilian, *Dromatherium*, but they are not dominant until the Eocene.

Man does not certainly appear in the geological record until after the glacial epoch.

Since we find in the Lower Cambrian rocks remains of such divergent groups as sponges, coralline cœlenterates, mollusca, trilobites, and brachiopods, and traces of chætopods, we are forced to conclude that if the evolutionary hypothesis is true, several earlier faunas must have preceded this. But it does not follow that we may anticipate the ultimate discovery of fossil remains of these, since it is not necessary to suppose that any skeletal parts were developed in these primordial forms. If they had no hard parts they would leave few, if any, fossil traces.

We will not pursue this subject further here, since the examination for which this is prepared scarcely covers even so much ground in this direction as this chapter. But we have given the summary here to fitly round off the general review of the material universe into which every work on physiography ultimately resolves itself.

No questions have been set upon the general distribution of life in the *Honours* papers since the title of the subject was altered from *Physical Geography* to *Physiography*, so that we do not append any to this section. But the matter in its broader relations is certainly within the extreme scope of the syllabus, as indeed a detailed account of the present geographical distribution of life would be. The student must use his individual judgment in the matter, though to know what is given here will do him no harm. If he would know more of the present distribution of life he will find an extremely interesting summary by Dr. Alfred Russel Wallace in the *Encyclopædia Brittanica* article, DISTRIBUTION.

INDEX.

LIST OF AUTHORS MENTIONED.